LABYRINTHS

Also by Catrine Clay

Trautmann's Journey:
From Hitler Youth to FA Cup Legend

King, Kaiser, Tsar:
Three Royal Cousins Who Led the World to War

Princess to Queen

Master Race:
The Lebensborn Experiment in Nazi Germany

LABYRINTHS

Emma Jung, Her Marriage to Carl,
and the Early Years of Psychoanalysis

CATRINE CLAY

HARPER
An Imprint of HarperCollins*Publishers*

HarperCollins books may be purchased for educational, business, or sales promotional use. For information, please email the Special Markets Department at SPsales@harpercollins.com.

First published in Great Britain by William Collins in 2016

FIRST U.S. EDITION

Library of Congress Cataloging-in-Publication Data

Names: Clay, Catrine, author.
Title: Labyrinths : Emma Jung, her marriage to Carl, and the early years of
    psychoanalysis / Catrine Clay.
Description: First edition. | New York, NY : Harper, 2016.
Identifiers: LCCN 2016013869| ISBN 9780062245120 (hardback) | ISBN
    9780062245151 (ebook)
Subjects: LCSH: Jung, Emma. | Jung, C. G. (Carl Gustav), 1875-1961—Marriage.
    | Psychoanalysts—Biography. | Psychoanalysis—History. | BISAC: BIOGRAPHY &
AUTOBIOGRAPHY / Social Scientists & Psychologists. | BIOGRAPHY &
AUTOBIOGRAPHY / Personal Memoirs. | PSYCHOLOGY / Movements / Jungian.
Classification: LCC BF109.J86 C53 2016 | DDC 150.19/540922 [B] --dc23 LC record available at
    https://lccn.loc.gov/2016013869

ISBN: 978-0-06-224512-0

16 17 18 19 20   OFF/RRD   10 9 8 7 6 5 4 3 2 1

For Gaby

Mini liebe Cousine und Helferin

Contents

'Their [marriage] partners can easily lose themselves in such a labyrinthine nature, sometimes in a not very agreeable way, since their sole occupation then consists in tracking the other through all the twists and turns of his character.'

Carl Jung

1

A Visit to Vienna

On Saturday 2 March 1907 Carl and Emma Jung arrived in Vienna for a five-day visit. They stayed at the Grand, the city's most fashionable hotel, just a few minutes' walk from the Opera and the famous Ringstrasse. Accompanying them was Ludwig Binswanger, an assistant at the Burghölzli lunatic asylum in Zürich, where Carl Jung worked as a doctor and first assistant to the director Eugen Bleuler. At ten the next morning the threesome waited to be collected by Sigmund Freud, who had invited them to Sunday luncheon at his family home, a short walk away at 19 Berggasse. None had met the Herr Professor before, though Jung and Freud had been corresponding for over a year.

Emma Jung was twenty-four, attractive – her wavy brown hair pinned up under a large hat – and very wealthy. But although her outfit was expensive, with its long skirts and furs against the March cold, it was not showy, because Emma herself was not showy. Carl was strikingly good-looking in a Teutonic sort of way – light brown hair, small moustache, dark eyes behind gold-rimmed spectacles, over six foot tall and powerfully built, with an imposing presence: a brilliant and ambitious young man just beginning to make his mark on the new and as

yet not very scientific field of 'Psychoanalysis', of which
Professor Freud, twenty years his senior, was already the
acknowledged master. Anyone observing Emma and Carl Jung
seated on the plush velvet canapé in the elegant foyer of the
Grand Hotel – with its chandeliers, ornate galleries, steam-
powered elevator and liveried footmen and porters – would
have seen a couple perfectly fitted to their surroundings: rich,
handsome, young, and, by all appearances, happy.

But appearances can be deceptive. Some time before the Jungs
left their home town of Zürich for Vienna, Emma had considered delivering her husband an ultimatum: either change his
ways or she would divorce him – a shocking and rare thing at
the beginning of the twentieth century, and one utterly alien to
the retiring character of Emma Jung. She felt lost in the labyrinth of her marriage to Carl, beset with problems. Life could
not go on the way it was.

By 1907 the Jungs had been married for four years and had
two young daughters – Agathe, aged three, and Gretli, almost
two – who were being cared for back in Zürich by the children's
maid, helped by Carl's mother and his unmarried sister Trudi.
Four years might not be long, but it was long enough for Emma
to discover the extent and complexity of her situation, though
not long enough to know what to do about it. The problem was
twofold: Carl's outward manner, so confident he could come
across as arrogant, concealed a very different and infinitely more
complicated interior, one which constantly eluded Emma,
however hard she tried to understand it. The second problem
was no easier to solve: Carl was always flirting with other
women, and they with him, provoking in Emma emotions she
had never even known she possessed: storms of jealousy and

fury followed by terrible feelings of self-doubt and recrimination. On top of all this Carl was extremely ambitious, working day and night at the asylum like a man possessed, driven by a conviction that he had a special understanding of the insane, because in many ways he was so like them. One way or another, it left almost no time for family life. Emma would spend hour upon hour in the Jungs' apartment on the second floor of the Burghölzli asylum waiting for her husband to return.

There were further complicating factors. Emma came from a well-known family of wealthy industrialists, the Rauschenbachs of Schaffhausen, making her one of the richest heiresses in Switzerland. Carl on the other hand was the son of a poor pastor of the Swiss Protestant Reformed Church. In fact, he was in debt when he first met Emma to the tune of 3,000 Swiss francs, which at a time when a working man might earn thirty francs a week was a very large sum indeed. It was a humiliation for Carl, this poverty, because the Jung family was respected in his home town of Basel and most of them were not poor at all. But his father, for reasons best known to himself, chose to work in remote parishes which barely offered a living. So they were the poor relations and Carl hated it. On marrying Emma he inherited all his wife's money and possessions, and became not only debt-free but a man of independent means in his own right, and able to support his impoverished mother and sister. Uncertain whether the young, ambitious Herr Doktor Jung was not just a plain old-fashioned fortune-hunter, Emma's family allotted her an additional monthly allowance of her own. In the event of a divorce everything each partner brought to the marriage reverted to them – a useful tool when delivering an ultimatum.

This is how things stood between the couple waiting for Herr Professor Freud in the foyer of the Grand Hotel in March 1907. One thing, however, was clear: Emma loved Carl fiercely and was prepared to fight to the bitter end to keep him. Carl, for his part, may or may not have married his wife for her money. But it certainly was not his only reason. The deeper reasons, true to his character, were infinitely more complicated.

The Jungs' spring break was a useful distraction, travelling to three countries over a period of three weeks, just the two of them, and staying at the very best hotels. Their tour would take them to Budapest, then Fiume and Abbazia in northern Italy, but their first stop was Vienna, where Carl had already made arrangements. 'I shall be in Vienna next Saturday evening, and hope I may call upon you on Sunday morning at 10 o'clock,' Carl had written to the *Hochverehrter* – highly esteemed – Professor Freud on 26 February. 'My wife has relieved me of all obligations while I am in Vienna,' he added, and 'I shall take leave, before my departure, to let you know at what hotel I am staying, so that you could if necessary send word there. Most truly yours, Dr Jung.'

Who knows what Freud made of Jung's repeated use of the first-person pronoun, as though he were coming on his own, but he must have understood because when he arrived at the Grand Hotel that March morning, Freud was bearing some flowers for Frau Doktor Jung. As he presented them to her, with a formal bow, Emma saw a small dapper man with a neatly clipped moustache and beard, dressed in his winter best: a worsted overcoat in the English style; a suit, waistcoat, floppy

4

bow tie, homburg hat, spats over boots, and a cane. His hair, inclined to be unruly, was combed down with his usual pomade supplied by the local barber he frequented every morning.

Originally the Jungs were meant to visit over Easter when Professor Freud had time to spare, but this was not possible for Carl and Emma who always spent Easter with their children. This presented a problem for Freud, who worked from eight in the morning till eight at night, apart from a break for the midday meal followed by a walk round the district for some air and a visit to the cigar shop. Unlike Carl Jung, Freud had not married into money and had to earn a living to support his large and extended family. 'But Sundays I am free, so I must ask you to arrange your visit to Vienna in such a way as to have a Sunday available for me,' he had written to Jung on 21 February. 'If possible I would like to introduce you to a small circle of followers on a Wednesday evening. I further assume you will be willing to forgo the theatre on the few evenings you will be spending in Vienna, and instead to dine with me and my family and spend the rest of the evening with me. I am looking forward to your acceptance and the announcement of your arrival.' Forgo the theatre? Jung had absolutely no intention of going. He hated the theatre. That was what Ludwig Binswanger was there for, to chaperone and entertain Emma. He himself wanted to spend every available minute with the Herr Professor.

As the four of them walked along, Freud smoking his third or fourth cigar of the day, the conversation remained on the formal level: the journey, the weather, their hotel accommodation – polite talk, just filling in. They must have made a comic sight – the two men who would become the two giants of psychoanalysis, walking side by side: Jung over six foot, Freud five foot seven, and Jung already talking his head off, loudly, as he always did.

Berggasse may have been only a short walk from the Ringstrasse, but it was far enough to leave the grandeur of the Ring district behind and enter the more modest, cobbled streets surrounding Vienna University. And it was long enough for a small misunderstanding to occur, seemingly insignificant at the time, but in hindsight very significant indeed. Freud, trying to put the visitors at their ease, joked that he was happy to be receiving them as guests to his home, but warned that it was a very modest place and had little to offer, just his *alte* (old woman) – nothing much else. Emma was shocked. So was Carl. What a way to talk about your wife, the mother of your children! Swiss haute bourgeoisie as they were, Freud's self-deprecating Jewish humour was completely lost on them. Emma, usually blessed with a healthy sense of fun, was puzzled.

Frau Professor Freud was there to greet the guests from Zürich as they mounted the stone steps of 19 Berggasse, and then up the stairs to the apartment. Emma, shaking Martha Freud formally by the hand – '*guten Tag, guten Tag*' – observed an older woman in a long-skirted, high-necked Sunday dress, stylish in its way, with a simple brooch at the neck, her dark hair pinned up neatly with combs. Behind her stood a younger woman, Martha's unmarried younger sister Minna who lived with them – an interesting set-up which caused plenty of gossip and speculation over the years. Herr Professor Freud and his two women. Apparently Minna sometimes answered the telephone calling herself 'Frau Professor'! Freud himself did not like the newfangled telephone and always left it to one of the women of the house to answer. In 1907 Sigmund Freud was fifty-one, Martha forty-six. They had been married for twenty-one years and had six children ranging from Mathilde twenty, Martin eighteen, Ernst sixteen, Oliver fifteen, Sophie fourteen, down to Anna, twelve. Freud called them his 'rabble' and his 'rascals' with affection and some pride.

Martha, greeting Emma Jung, observed a formal, reticent young woman, dressed with an understated style, and only a few years older than her daughter Mathilde. She already knew about the Frau Doktor's wealth. Everyone knew. But what she could not fail to notice was that Frau Doktor Jung did not act wealthy. As the two women exchanged greetings, Jung was shaking hands in the formal Swiss way. No fancy Viennese kissing of ladies' hands for Carl.

On special occasions such as this Martha would cook a chicken for the midday meal, though her husband was not keen. 'Let them live,' he always said. 'Let them lay eggs.' But his wife made all the household decisions, it had been that way from the start: he earned the money, she ran the house. It suited them both and Freud never interfered and never complained. So there they sat, twelve of them, round the dinner table having their meal, served from the kitchen by the maid, with the men at one end and the women and children at the other. Emma was surprised by all the talk – the Freud children expressed themselves easily and cheerfully. 'Our upbringing might be called liberal,' Martin Freud later wrote, reflecting on his 'gay and generous' father. 'We were never ordered to do this, or not to do that; we were never told not to ask questions.' For their part, Martha and Minna, the two sisters, could not fail to notice that the Jungs both had strong Swiss accents, making them seem more provincial than they were. Apparently, Frau Doktor Jung had wanted to attend

Zürich University to study the natural sciences but her father wouldn't let her. Understandably. The only women who attended university were rich foreigners – Russians and the like.

Looking back, Martin Freud, as an observant eighteen-year-old, remembers Herr Doktor Jung having a 'commanding presence': 'He was very tall and broad-shouldered, holding himself more like a soldier than a man of science and medicine. His head was purely Teutonic with a strong chin, a small moustache, blue eyes, and thin closely cropped hair.' In fact, Jung's eyes were brown not blue, but in the almost entirely Jewish environment inhabited by the Freud family the Teutonic aspect was unusual and rather interesting. Still, Martin took a dislike to Jung: 'He never made the slightest attempt to make polite conversation with mother or us children but pursued the debate which had been interrupted by the call to luncheon. Jung on these occasions did all the talking and father with unconcealed delight did all the listening,' he wrote, still irked by the memory. Martin was doubly surprised because normally his father strongly disapproved of visitors ignoring his family and he would deliberately change the conversation to include them, making it clear that this was not how things went in the Freud household. But not with Herr Doktor Jung, who talked throughout the meal exclusively to his father, showing no awareness of or concern for anyone else. Emma might have told them this was typical of her husband, who had gained some enemies amongst his colleagues at the Burghölzli asylum because of it, though never amongst his patients who all revered him.

After the meal there was coffee and then Emma and Binswanger took their leave, as arranged, so Carl could spend more time with Freud. The two men swiftly retired to Freud's consulting rooms on the mezzanine floor below, which looked out over a small garden with a single chestnut tree. They talked for thirteen hours without a break. It was love at first sight, a mutual enchantment accompanied by every high hope. Freud had read this brilliant young doctor's papers on *dementia praecox* – or 'schizophrenia' as Bleuler of the Burghölzli asylum had coined it – as well as Jung's 'Experiments in Word Association', and now, to his delight, he found that the man in conversation was as brilliant and challenging as the writer on the page.

As for Carl Jung's first impressions of Freud: 'In my experience, up to that time, no one else could compare with him,' he wrote later. 'There was nothing the least trivial in his attitude. I found him extremely intelligent, shrewd, and altogether remarkable.' Carl did not get back to the hotel until two in the morning, having to call out the night porter, by which time Emma was fast asleep.

Over the next five days a routine was established: the visitors would be picked up from the hotel every morning by one of the Freud family to be shown around the city. By the end of each day's sightseeing everyone was exhausted, everyone except Jung, who enjoyed a rude energy throughout his life, and who would hurry along to 19 Berggasse for his late-night sessions with Freud, talking psychoanalysis, the new and shocking movement that the Herr Professor was leading with missionary zeal, and which, it soon transpired, he meant to bequeath to this brilliant young doctor from the Burghölzli asylum, naming him his 'crown prince and heir' with typical impulsiveness, and much to the annoyance of his Viennese colleagues.

The reasons were obvious: not only was Carl Jung brilliant, young and energetic, he was charismatic – an essential prerequisite for a leader. In addition, all the other men in the Viennese group were Jewish, whereas Jung was a Gentile, an Aryan from Switzerland. Freud knew this was the only way psychoanalysis could reach a wider, international public, transforming it into a world movement. He knew it because he had lived with anti-Semitism all his life. As much as he tried to ignore it he knew he could never overcome it. As he wrote to one of his most loyal followers, Karl Abraham, in December 1908: 'Our Aryan comrades are really completely indispensable to us, otherwise Psycho-Analysis would succumb to anti-Semitism.'

The Vienna Emma and Carl Jung visited at the beginning of the twentieth century was a great cosmopolitan city of 2 million inhabitants, only half of whom had *Heimatberechtigung*, that is, were legally domiciled Viennese German-speaking Austrians. The rest came from the four corners of the Austro-Hungarian Empire: Bohemians, Moravians, Hungarians, Poles, Czechs, Croats, all bringing different languages and embracing different religions. And Jews, many Jews. These ranged from Vienna's poorest inhabitants living in the slums of Leopoldstadt to the new professional classes, lawyers, writers, journalists, artists and doctors like Freud, and the very richest: the fabulously wealthy merchants and bankers who lived in the nouveau riche Ring district in houses so large they were referred to as *palais*, often built in the neo-Renaissance style with columns, loggias and caryatids. It was these wealthy Jews who had helped finance Emperor Franz Joseph's transformation of Vienna from the walled medieval city it once was to the capital of imperial grandeur which Emma and Carl saw all about them.

The sheer scale of it all was staggering. The grand boulevard of the Ring offered a dramatic setting for the Rathaus and

Reichsrat, the city hall and parliament, as well as the Opera, the Burgtheater, the churches of St Stephen and the Votivkirche, the stock exchange, and, leading to the Heldenplatz, a vast columned piazza in front of Kaiser Franz Joseph's Hofburg palace adorned with two massive equestrian statues, one of Prince Eugene of Savoy, the other of Archduke Charles of Austria. Then there were the museums – the one dedicated to natural history being of particular interest to Emma – and the many parks where you could wander up statue-lined avenues, sit by fountains or listen to one of the military bands playing Viennese waltzes and marches or melodies from the latest operetta. Everywhere you looked there were uniforms, army officers of the empire in red or pale blue, with sashes, epaulettes, gold braid, plumed helmets, swords, sabres, and highly polished boots. Every official appeared to have a uniform too, even the tram drivers, and little boys were often dressed in miniature military uniforms for their Sunday best.

The Hungarian court put on frequent displays of imperial pomp and power, such as the City Regiment's daily march. On one occasion when Martin Freud was with Carl and Emma in the Ring district, Emperor Franz Joseph's coach drove past, resplendent in red and gold with liveried coachmen and postilion. The Jungs had never seen such a thing and Martin was amazed to see the Herr Doktor, usually so superior, pushing to the front of the crowd, 'like a small boy' thrilled to catch a glimpse of the emperor with his companion, the former actress Katharine Schratt, seated at his side. For the Jungs, from small, republican Switzerland, it was the stuff of fairy tales. Whether visiting the famed Schiffmann's department store, illuminated with the latest forms of electric lighting, or Demel's, where the cakes and the *Sachertorte* were the best in the world, or joining the daily *Corso* along the Kärntner Ring, where Viennese society paraded in the latest fashions, everything dazzled. The Baedeker travel guide put it nicely: 'With limited time, a week would suffice for a superficial overview of everything worth seeing.'

In the evenings after Carl had left for 19 Berggasse, Emma rested in the hotel before venturing out into the city – to the Burg Theatre or the Opera or to one of the famous Viennese operettas, perhaps with Binswanger, perhaps with one of the Freud ladies. Or she might stay in the hotel and dine in their *Salle de Diner* and later sit in one of their more intimate *salons* to read before retiring to her bedroom. She was surrounded by opulence, the Grand being the very first of the fashionable hotels of Vienna, built in the 1870s, with 300 bedrooms, half of which had ensuite bathrooms – a luxury as yet unheard of in Zürich. It had central heating and electric lighting, its own telegraph office in the foyer from which guests could send telegrams and make telephone calls with the assistance of well-trained telegraphists, and a private fiacre carriage service to take them anywhere they wished. So Emma could not have been better cared for whilst her husband was with Professor Freud. But she might have been happier if Carl had whiled away some of those evening hours with her.

Aside from psychoanalysis and the campaign to conquer the world, Freud and Jung talked about themselves – a natural transition since psychoanalysis dealt with neuroses and psychoses and all manner of obsessive behaviours, most of which appeared to have their roots in childhood, including their own. The subject they discussed most was sex: specifically, Freud's theory that sexual trauma in childhood was the root cause of later neuroses and hysteria. It might be sexual abuse by a stranger or a family friend, or by a family member in which case it was incest. Freud had many examples from his own patients who came to him in the first instance because they were unaccountably paralysed, or suffering from chronic anxiety, depression, physical pain, sleeplessness, paranoias. Time and again it transpired that they were repressing early sexual experiences, though by 1907 Freud had modified his earlier view that all cases of hysteria had a sexual origin. He and his colleague and teacher Professor Breuer had published their *Studies on Hysteria* in 1895, by which time some doctors were diagnosing their female hysterics with sexual dysfunction and treating them with hypnosis or various forms of massage, including that of their genitals to bring about orgasm. But these were not subjects spoken about openly, except by Freud, who made sexual repression the linchpin of his work, shocking the general public and plenty of his medical colleagues in the process. His 'cure' was revolutionary: the 'talking cure' of psychoanalysis, designed to uncover the origin of the neuroses rather than merely treating them. The unconscious, Jung and Freud agreed, was the key to everything. And the key to the unconscious was the dream.

Carl had been conducting a good deal of research into the unconscious himself, especially the 'Word Association' tests he carried out in his laboratory at the Burghölzli on both 'normal' and 'abnormal' subjects – using a galvanometer to measure the patients' reactions by applying a weak electric current to the subject which measured the fluctuations in the skin with each association, the reaction recorded on a graph. He had also read Freud's most famous work *The Interpretation of Dreams*, published in 1900, which boldly stated: 'In the following pages, I shall demonstrate that there exists a psychological technique by which dreams may be interpreted and that upon the application of this method every dream will show itself to be a senseful psychological structure which may be introduced into an assignable place in the psychic activity of the waking state.' So the two men were in agreement about the powers of the unconscious. What Carl Jung could not agree with, however, even before he met Freud in person, was the central role played by childhood sexual trauma. 'It seems to me,' he wrote to Freud on 5 October 1906, six months before his visit to Vienna, 'that though the genesis of hysteria is predominantly, it is not exclusively, sexual.' He thought this might be because 'I: my material is totally different from yours [working mostly with uneducated insane patients], II: my upbringing, my milieu, and my scientific premises are in any case utterly different from your own, III: my experience compared to yours is extremely small.' Freud, superior in age and experience, was happy to bide his time. Sooner or later the crown prince would see the light. He could not know that Jung had personal as well as professional reasons for believing what he did.

As Freud had mentioned in his letter to Jung, on Wednesdays there was always a meeting at 19 Berggasse of his Viennese colleagues, including Alfred Adler, Rudolf Reitler, Max Kahane and Wilhelm Stekel – later joined by Paul Federn and Eduard Hitschmann and occasionally Sandor Ferenczi and Otto Rank. On that particular Wednesday in March 1907, Binswanger, attending with Carl Jung, remembered there were only five or six others present.

They assembled in Freud's consulting room as usual – a room filled with cigar, pipe and cigarette smoke, dimly lit with gas lamps and surrounded by Freud's growing collection of antique and oriental art – with wine brought to them by Martha Freud. The atmosphere was relaxed, with no etiquette and plenty of humour. Binswanger was amazed how Freud could dominate the evening so completely after a long day's work. The subject was exclusively psychoanalysis: the interpretation of dreams, neuroses, paranoias, childhood sexuality and so on – Freud offering detailed examples from his cases, then listening care-fully, answering questions with gestures of the hand, sometimes holding one of his smaller antiques, and always the cigar. His manner was simple and filled with charm, Binswanger recalled, but you could never forget you were in the presence of greatness.

Jung took part in the discussion, but less loudly than usual. The evening left him perplexed. 'I felt so foreign before this Jewish intellectual society. That was something completely new to me. I had never experienced that before,' he told his friend Kurt Eissler some years later. 'I found it very difficult to adjust, to adopt the right tone.' Their conversation had 'a certain cyni-cism', he said, and it made him feel 'like a country bumpkin'. When Freud joked: 'You wouldn't be an anti-Semite now, would you?' Jung took it seriously, answering: 'No, no. Anti-Semitism is out of the question,' which must have amused Freud quite a bit.

At the end of the evening the Herr Professor turned to his Swiss guests and said: 'Now you've seen the whole *Bande*, the whole gang.' That shocked Binswanger; it seemed so dismissive. Something else shocked him too. Earlier in the week he, Freud and Carl Jung had been analysing one another's dreams. In later life he recalled Freud's interpretation of Jung's dream, though not the dream itself: apparently Jung had a hidden wish to dethrone Herr Professor Freud and place the 'Crown of Psycho-Analysis' on his own head. It was that Jewish–Viennese humour again. Binswanger's own dream was of arriving at 19 Berggasse to find it was the old, not the newly renovated entrance, with two ancient gas lamps hanging outside. That, announced Freud, revealed Binswanger's wish to marry Mathilde, his eldest daughter – and then his deciding against it because the place was too shabby. But it was said with a laugh, and even Binswanger got the joke.

When the five-day visit was up the Jungs took their leave with many heartfelt thanks, and travelled on to Budapest by the Continental Express sleeper, in the comfort of a first-class carriage. Binswanger stayed behind for another week with Freud, establishing a friendship that would last a lifetime. On the train, looking out at the passing hamlets and farms and the Alps beyond, Emma had no idea how deeply her husband had been affected by his encounter with Freud. And if she had, she could hardly have guessed the reason why.

The idea of visiting Budapest originally came from Emma's wish to see the city where her father had established a branch of the Rauschenbach family business in agricultural machinery. She wanted to go to the premises and perhaps meet some of the employees her father had known when he was there back in the

1880s. But if Emma hoped for some time sightseeing with Carl she was mistaken. He showed little interest in it, preferring to spend his time with a colleague, Philip Stein, discussing medical cases and his own recent experiments in word association, so Emma was forced to do most of her sightseeing on her own.

Worse was to come when they arrived in Abbazia, a fashionable resort on the Adriatic coast. A woman staying at their hotel struck up a conversation with them at dinner on their first evening. She was attractive, intelligent, Jewish – a lady of independent means with progressive opinions and quite fascinated by the new and daring science of psychoanalysis. And even more fascinated by the charismatic and handsome Herr Doktor, as most women were. And Carl, in the wake of his infatuation with Freud and all things Jewish, was a willing partner. Every evening he and the woman retreated to a sofa in the corner of the drawing room to discuss psychoanalysis. If Emma joined them the woman talked down to her as the mere wife. Emma, jealous and humiliated, complained to Carl, but he told her there was nothing to it – their discussions were purely professional. It took him two years before he could admit the truth: that this was another of his 'infatuations'.

2

Two Childhoods

Emma Rauschenbach first met Carl properly when she was seventeen. She had just returned home to Schaffhausen in eastern Switzerland from Paris where she had been staying with friends of the family, being 'finished off' in preparation for marriage to a suitable young man from a similar haut-bourgeois Swiss background to her own. She was shy and quiet, but clever, always top of her class at the *Mädchenrealschule*, the local school for girls from every kind of background, rich and poor alike. She had not wanted to go to Paris; she had wanted to continue her education and go to university to study the natural sciences, a subject which had fascinated her since childhood, but it was not considered the right path for young Swiss women like Emma, and her father would not hear of it. Instead she went to Paris to perfect her French and acquaint herself with La Civilisation Française. Serious young woman that she was, Emma spent hour upon hour in the museums and began to learn Old French and Provençal in order to read the legend of the Holy Grail in the original – the twelfth-century romance about Perceval, a knight in the Arthurian legends, that would fascinate her for the rest of her life. By the time Carl Jung came to pay a visit she was informally engaged to the son of one of her father's wealthy Schaffhausen business colleagues, and her future lay predictably before her.

Emma's childhood home, the Haus zum Rosengarten (the House of the Rose Garden), was an elegant seventeenth-century mansion situated on the banks of the Rhine. It had been bought by Emma's grandfather Johannes Rauschenbach with the fortune he made from his factory producing agricultural machinery, exported worldwide, and the iron foundry next to it, both within walking distance of the house. Later he augmented his fortune by buying the Internazionale Uhren Fabrik (the International Watch Company, IWC), an American firm producing the first machine-made fob and wrist watches. Emma's grandfather died young in 1881 and her father Jean, aged twenty-five, took over the running of both factories and moved into the house, still lived in by his mother, with his young wife Bertha. Their daughters, Emma and Marguerite, were both born there: Emma on 30 March 1882, and Marguerite fifteen months later.

Emma recalled her childhood as being idyllic, combining untroubled happiness and privilege in equal parts. Her nickname was 'Sunny' and her life at that time gave her no reason to feel otherwise. The house itself, large, square, solid, was separated from the banks of the Rhine by a formal rose garden, laid out by her uncle Evariste Mertens, a landscape designer, and which gave the house its name. Schaffhausen itself was a prosperous town of fine Renaissance buildings with stuccoed and frescoed façades adorned with high-minded words exhorting the good burghers to lead virtuous lives. In the back streets and away from the grandeur stood the many factories, small industries and workshops which were the foundation of its wealth, a tribute to the Swiss tradition of hard work, and to the benefits of hydroelectricity, derived from the power of the massive Rhine Falls nearby. 'Standing in the window,' recalled Gertrud Henne, Emma and Marguerite's cousin who came to the house to play with the sisters, 'I liked to watch the big "Transmissions": pillars standing in the Rhine with giant wheels that conducted hydropower via cables to the various factories along the Rhine.' Anyone with ambition might make themselves a fortune in those heady early industrial days in Schaffhausen, and Johannes Rauschenbach, who started with nothing more than a machine repair shop, then a pin factory supplying the local cotton industry, and finally the world-renowned agricultural machinery factory, became the wealthiest of them all, and one of the richest men in Switzerland.

When Jean Rauschenbach took over the business, with factories at home and abroad, Emma's mother and grandmother took over the running of the house. Grossmutter Barbara lived in rooms upstairs and liked to sit in a fauteuil by the window overlooking the Rhine, reading her *Gazette* with her lorgnon, wearing a large bonnet with ribbons and surrounded by her collection of dolls, kept in a large old wall bed, and which the girls were sometimes allowed to play with. Having started life modestly, Grossmutter Barbara never fully accustomed herself to the great wealth the family came to enjoy. 'If only you'd remained a mechanic,' she used to tell her husband.

The two sisters were very different but they were close and remained so all their lives. Emma could spend hours on her own, reading, writing, thinking. Marguerite was less the thinker, more the sporty, outward-going type, and moodier. Both sisters played the piano well, but Marguerite liked to sing too, and play-act, and she swam in the Rhine in all weathers, right into old age. They shared a private tutor before moving up to the local school for girls, and their upbringing was conventional Swiss haut bourgeois, instilling the values of a Protestant work ethic, social conformity, and feminine grace and good manners, so they knew how to behave when Herr Direktor Rauschenbach and his wife gave one of their grand receptions required of the foremost family of Schaffhausen.

Emma at school, third row, third from right.

Both girls adored their mother, Bertha, who allowed her daughters plenty of freedom. For this the Haus zum Rosengarten was perfect, with its large cobbled courtyard, extensive outhouses, and the stables where the girls kept their horses, Lori and Ceda, looked after by Reeper, the groom, an ex-cavalry officer in the Austro-Hungarian Army. If it was raining there were plenty of toys to play with inside; if it was snowing there was sledging and ice-skating; if it was one of those heatwave summers there was swimming in the Rhine, and all year round Reeper took them out riding to villages and castles and other local landmarks.

The question of Carl and Emma's first meeting is a moot one: was it in 1896 when he was still a student or was it three years later, when he was poised to take his first job working as a lowly assistant physician at the Burghölzli asylum and Emma had just returned from Paris? If it was in 1896, then it was at the Haus zum Rosengarten and it was an event hardly even remembered by Emma. But if it was in 1899 it was at Ölberg, the Mount of Olives, an ancient property like a small castle, square and thick-walled, with its own medieval chapel, the St Wolfgangs Kapelle, high on the slopes overlooking Schaffhausen with a drive so long and steep you could not see from one end to the other. The family had spent every summer there since the girls were small. But by 1899 Jean Rauschenbach had decided to sell the Haus zum Rosengarten and make Ölberg their family home, replacing the beautiful little castle with a *Jugendstil* mansion, a vast stone pile in the heavily ornate style of the time, with turrets and gables and oriels, high-ceilinged reception rooms, and a grand stairway leading up to a wide landing with bedrooms and bathrooms off.

One entire floor was set aside for Emma and Marguerite, then in their teens, and the whole house was lit by electricity, heated by central heating, and served by a raft of servants inside and out. The architect was Ernst Jung of Winterthur, by chance one of Carl Jung's uncles, who had already renovated the Sonnenburg property next door which belonged to Emma's landscape architect uncle Evariste Mertens, who now proceeded to design

22

the far grander gardens at Ölberg. According to his own account, as soon as Carl first set eyes on Emma, in the half-light, coming down the grand stairway into the hall, he decided this was the girl he would marry. If it was 1899 then Emma would have been seventeen, just back from Paris, more self-assured than before but still shy and retiring, poised on the edge of adulthood. If it was 1896, as Carl described in *Memories, Dreams, Reflections*, that would make Emma just fourteen.

How an impoverished medical student came to be visiting this prominent and unimaginably wealthy family in the first place is down to Emma's mother. Bertha was the beautiful daughter of Schenk, patron of the local *Gasthof*, a successful family business, providing rooms as well as excellent food – but still a *Gasthof*. So when Bertha married Jean, the son and heir to the Rauschenbach fortune, she married well above her station. But, rather like her mother-in-law, she never forgot her humble origins. Bertha knew the Jung family because she and Carl's mother, Emilie Preiswerk, had attended the same school, and the Schenk *Gasthof* was in Uhwiesen, one of three villages in the parish of Laufen by the Rhine Falls where Carl's father was pastor. The living at Laufen was poor: only enough to employ one maid-of-all-work, which included looking after infant Carl when his mother was not 'well', which was often. Bertha Schenk was one of Pastor Jung's parishioners and she helped him out from time to time, taking the baby for walks along the Rhine in his pram. Years later Carl still remembered her as she was then: 'the young, very pretty and charming girl with blue eyes and fair hair' who 'admired my father'.

Now, encouraged by his mother, who had remained in touch with Bertha, he decided to pay Frau Rauschenbach a visit, and there saw the daughter Emma coming down the stairs. Even if Carl first clapped eyes on Emma when she was fourteen, she herself first knowingly met Carl in 1899 when she was seventeen. And the first correspondence between them dates from this year, when Emma returned from Paris. It was a one-way sort, that is, mostly from Carl, starting with picture postcards addressed formally to '*Sehr geehrtes Fräulein!*' – most esteemed young lady, always ending in an exclamation mark.

It took Carl many months before he could summon up the courage to ask Emma to marry him, and when he did she refused him. 'For various reasons I was turned down when I first proposed,' he wrote to Freud in 1906: 'later I was accepted, and I married.' Various reasons, plural. One was that Emma was already engaged, albeit informally, to the son of one of her father's business colleagues. Another: Carl's loud and rumbustious personality was utterly overwhelming for a young woman such as Emma. Another: Carl Jung did not have a penny to his name, nor was he ever likely to have, since by then he had decided to be an *Irrenarzt* – a doctor of the insane – the most lowly of professions. This presented a serious social barrier and was so shocking that Emma's father could not be told. Their engagement, when it finally occurred, was a secret one.

What changed Emma's mind? The short answer is probably Carl himself. It took him a while but he was utterly determined and marshalled everything he had to win her hand, starting with certain natural advantages: his good looks, his imposing presence, his challenging conversation, his intelligence, his lively humour, and what he himself called his 'intuition'.

Carl's intuition told him that beneath her reticent, formal manner Emma was yearning for something less conventional, more intellectually satisfying, more adventurous – an outlet for her cleverness which she could not have if she married her haut-bourgeois beau. So he embarked on his campaign, bombarding her with letters filled with fascinating ideas and amusing self-deprecating comments. He told her about his favourite writers and philosophers, his love of mythology, his work, and he confided in her about his ambitions, his hopes and his fears. And he gave her lists of books to read for discussion next time they met. A seduction by intellect.

Even so, Emma still refused him. Everything about Carl, his physical size, his huge personality, his brilliance, was too power-ful for her. How was she to know that Carl had another self, well hidden, full of doubt and complexes and feelings of social inferiority? One refusal was enough for 'other Carl'. 'Father would never have asked her again,' their son Franz confirmed years later. 'He was crushed. He was poor, and not on the same social level as Emma, and so he thought he didn't have a chance.' Carl thanked Emma for her honesty and withdrew. These were his early months working at Burghölzli asylum and he became so plagued with insecurity that he hid himself behind the high walls of the institution. By his own account, he never went out for six months, causing colleagues to think he was behaving more like an inmate than a doctor. As for Emma: 'My mother was very shy then, and introverted,' said Franz. 'She was afraid to move ahead, to say yes.'

But Carl had a key ally within the Rauschenbach family. Intelligent, modern in outlook, and coming from modest begin-nings herself, Emma's mother Bertha saw nothing wrong with Carl Jung, the lowly assistant physician now employed in a lunatic asylum. Money? Emma had plenty of money. Bertha remembered the little boy she had pushed alongside the Rhine in his pram, now grown into a fine young man, and here was her daughter Emma, the clever, studious one – and what did it matter that she was engaged to another young man? It was hardly an engagement at all, nothing fixed, nothing formal.

Emma adored her mother and without her encouragement she would probably never have found the courage to marry Carl. After some months Frau Rauschenbach contacted Carl, arranged to meet him in a restaurant in Zürich, and urged him not to be put off and try asking Emma once more. She even invited him back to Ölberg, sending her own green carriage and coachman to collect him from Schaffhausen station. And this time, in October 1901, Emma said yes. Once decided, she never wavered. He need not worry, she assured him: she knew exactly what she was doing.

But Emma said yes to the Carl she knew: the extrovert, clever, handsome Carl with his earthy energy and loud exuberant laugh, not the 'other' Carl, the hidden one. Had she known the strangeness and complexity of the 'other' Carl – had she been able to see what lay ahead – she might have answered differently. Or not. Over the weeks till their secret engagement she caught glimpses of this 'other' Carl. To her surprise, it was she who had to reassure him, again and again, of her love. She thought it would be the other way round.

'My situation is mirrored in my dreams,' Carl wrote in his 'secret diary' in December 1898, whilst still a medical student:

Often glorious, portentous glimpses of flowery landscapes, infinite blue seas, sunny coasts, but often too, images of unknown roads shrouded in night, of friends who take leave of me to stride towards a brighter fate, of myself alone on barren paths facing impenetrable darkness. 'Oh fling yourself into a positive faith,' my grandfather Jung writes. Yes, I would be glad to fling myself if I could, if that depended only on the uppermost me. But an inexplicable heavy something, a listlessness and numbness, weariness and weakness, always prevents the final step. I have already taken many steps, but I am still a long way from the final one. The greater the certainty, the more superhuman the doubts . . .

This was and always would be the crux of the matter for Carl: he had a personality which was split: sure and unsure, optimistic and pessimistic, introverted and extroverted, sensitive and insensitive, brilliant yet obtuse; genial yet given to violent rages; cold under warm, dark under light – always split, and that split always hidden. Secret.

Later he called them 'Personality No. 1' and 'Personality No. 2', but growing up he hardly knew the difference. At the parsonage of Klein-Hüningen near Basel, where the Jung family moved when Carl was five, there was an old wall in the garden made of large blocks of stone and in the gaps between the stones he lit small fires which had an 'unmistakable aura of sanctity' about them and had to burn 'for ever'. One stone jutted out of the wall. 'My stone,' he called it:

Often, when I was alone, I sat down on this stone, and then began an imaginary game that went something like this: 'I am sitting on top of this stone and it is underneath.' But the stone could also say 'I' and think: 'I am lying here on this slope and he is sitting on top of me.' The question then arose: 'Am I the one who is sitting on the stone, or am I the stone on which *he* is sitting?' This question always perplexed me, and I would stand up, wondering who was what now. The answer remained totally unclear, and my uncertainty was accompanied by a feeling of curious and fascinating darkness. But there was no doubt whatsoever that this stone stood in some secret relationship to me. I could sit on it for hours, fascinated by the puzzle it set me.

He sat on that stone for hours, trying to work out whether it was him, or he was it. This was how Carl Jung described it to his assistant, the analyst Aniela Jaffé, when he was eighty and finally agreed to recount his life. Thinking about it then, he added:

> Thirty years later I again stood on that slope. I was a married man, had children, a house, a place in the world, and a head full of ideas and plans, and suddenly I was again the child who had kindled that fire full of secret significance and sat down on a stone without knowing whether it was I or I it. I thought suddenly of my life in Zürich, and it seemed alien to me, like news from some remote world and time. This was frightening, for the world of my childhood in which I had just become absorbed was *eternal*, and I had been wrenched away from it and had fallen into a time that continued to roll onwards, moving further and further away. The pull of that other world was so strong that I had to tear myself away violently from the spot in order not to lose hold of my future.

A strange child and a strange childhood. When the young Bertha Schenk came to take the infant Carl for walks along the Rhine,

Pastor Jung had often been looking after his son on his own, the mother, Emilie, being 'away' in some unknown place for people suffering from unknown ills. 'Dim intimations of trouble in my parents' marriage hovered around me,' Carl later recalled. As a child he fell ill with fever and suffered horribly from eczema. 'My illness, in 1878, must have been connected with a temporary separation of my parents. My mother spent several months in a hospital in Basel, and presumably her illness had something to do with the difficulty in the marriage.' Usually the maid looked after him, but often it was his father. 'I was deeply troubled by my mother's being away. From then on, I always felt mistrustful when the word "love" was spoken.' Whilst Emilie was 'away' Carl slept in his father's room. He remembered his father carrying him in his arms, trying to get him to sleep, pacing up and down, singing his old fraternity student songs. When Carl's mother finally came back home his parents no longer shared a bedroom. Frightening things emanated from her room, indefinite figures, floating, headless, luminous. Carl had 'vague fears' and heard strange things in the night, all mixed up with the muted roar of the Rhine Falls nearby. He could not breathe and thought he would suffocate. 'I see this as a psychogenic factor,' he later told Aniela Jaffé; 'the atmosphere of the house was beginning to be unbearable.' He went on sleeping in his father's room throughout childhood. In fact, until he was eighteen and preparing to go to Basel University.

'I had never come across such an asocial monster before,' recalled Albert Oeri, one of Carl's few playmates during those early years. Albert had been brought to the parsonage by his father, an old student friend of Pastor Jung's, to play with Carl. 'But nothing could be done. Carl sat in the middle of the room, occupied himself with a little bowling game, and didn't pay the slightest attention to me.' Carl was not used to playing with other children, not even the village children who were anyway mostly out in the fields helping their parents with haymaking or herding the cows. When the Jung family moved to Klein-Hüningen, Albert's family still sometimes visited on a Sunday afternoon. By now a different Carl had made an appearance: extrovert Carl, boisterous Carl, the one who did not like weaklings, especially one of his cousins whom he teased mercilessly. 'He asked this boy to sit down on a bench in the entrance way,' recalled Oeri. 'When the boy complied, Carl burst into whoops of wild Red Indian laughter, an art he retained all his life. The sole reason for his satisfaction was that an old souse had been sitting on the bench a short time before and Carl hoped that his sissy cousin would thus stink of a little schnapps.' But the moment he had done it he regretted it. Introvert Carl did not want to hurt anyone.

Under the loud whooping lurked the other Carl, the one with secrets to hide. The first of these was a dream he had when he was four, one so significant and so terrible he never told anyone about it until he was sixty-five: 'A dream which was to preoccupy me all my life.' He was in a meadow when he discovered a dark hole which he had never seen before, stone-lined, with a stone stairway leading far down. Fearfully he descended. At the bottom there was a doorway with a round arch and a heavy green curtain, brocade, leading through to a rectangular chamber with an arched ceiling, again of stone. A blood-red carpet ran from the entrance to a low platform on which stood a golden throne and on this throne stood something which he first took to be a tree trunk, twelve to fifteen feet high and two feet thick: a huge thing, reaching almost to the ceiling and made, he then realised, of skin and naked flesh. On the top was a rounded head with no face or hair, only a single eye. An aura of brightness wafted above it. Carl was paralysed with terror, believing it might at any moment crawl off the throne like a worm towards him. At that moment he heard his mother calling from above: 'Yes, just look at him. That is the man-eater!' and he woke sweating and scared to death. 'This dream haunted me for years.' Much later he realised it was an anatomically accurate phallus.

When the family moved to the old parsonage at Klein-Hüningen, Pastor Jung became chaplain at the local lunatic asylum as an additional role and Carl started going to school. Academic work was easy for him, but not the social side – he was not used to other children and they were not used to a child as strange as him. In time he learned to join in but he always felt it alienated him from his true self. At home he played alone for hours, hating to be watched, building high towers with wooden bricks, making drawings of battles and sieges, lighting fires in the garden. When he was ten he did something which was totally incomprehensible to him, even then: he had a ruler in his pencil case, of yellow unvarnished wood, and out of it he carved 'a little manikin about 2" long, with a frock coat, top hat, and shiny black boots. I coloured him black with ink, sawed him off the ruler, and put him in the pencil case, where I made him a little bed. I even made a coat for him out of a bit of wool.'

He also put a smooth blackish stone from the Rhine in the pencil case, painted to divide it into an upper and lower half. This was *his* stone. 'All this was a great secret. Secretly I took the case to the forbidden attic at the top of the house [forbidden because the floorboards were worm-eaten and rotten] and hid it with great satisfaction on one of the beams under the roof – for no one must ever see it! I knew that not a soul would ever find it there. No one could discover my secret and destroy it.' He used to go up there to visit the manikin, always surreptitiously, and deposit tiny scrolls in the pencil box for him, written in a secret language. Like sitting on the stone, it always made him feel better, bringing him back to his true self. This ritual lasted for about a year. Then he forgot all about it till he was thirty-five and writing *Wandlungen und Symbole der Libido*, later translated as *Transformations and Symbols of the Libido*, the book which would signify his final break with Sigmund Freud.

Apart from arithmetic, which always remained a terrifying mystery to him, Carl was clever and when he was eleven he easily gained a place at the *Gymnasium*, the grammar school, which was situated in the precincts of Basel cathedral. The work, classics-based, was no problem – he already knew Latin which his father had taught him since the age of six, and he was already widely read, especially the Bible. If anything, boredom was the problem. But social life was another matter. Here came Carl, the poor parson's son, walking from his village far out in the countryside, through meadows and woods and fields, in his bumpkin clothes and holes in his shoes so he had to sit for the rest of the school day in wet socks, talking in his broad yokel Basel dialect. And there came the well-dressed sons of the foremost families of Basel in horse-drawn carriages, with fine manners, plenty of pocket money, talking in refined High German or French about their holidays in the Alps, and Carl, having no holidays, felt an envy he had never felt amongst the poor farmers' sons who had been his classmates at his local school.

Now, for the first time, he realised that his family was poor, and when any of his classmates invited him to their grand houses he felt 'as timid and craven as a stray dog'. His feelings of inferiority, fatefully accompanied by equally powerful feelings of superiority, were exposed to the world: 'My shoes are filthy, so are my hands; I have no handkerchief and my neck is black with dirt.' His first year was completely ruined, he said, because he had the 'disagreeable, rather uncanny feeling' that he had 'repulsive traits' which caused the teachers and pupils to shun him, and it is true – many pupils did shun him, even at times Albert Oeri who was in the same class, because Carl was just too strange, too uncouth, too different. The only boys he spent his time with, if at all, were the sons of farmers, the poor ones who spoke the same local dialect. It did not help that he was clever, thirsty for knowledge, arrogant. On one occasion a teacher accused him of cheating because he could not believe this boy could write such an essay on his own. Carl was mortified. He had spent hours of hard work on it. Grown big by now, he got into plenty of fights and brawls. But he always felt 'a certain physical timidity' – a feeling that he was somehow repulsive.

In his twelfth year he had what appears to have been a breakdown. As he described it, he was standing in Basel cathedral precinct one day in early summer, waiting for a classmate before setting off on the long trek home, when another boy from the *Gymnasium* knocked him over and as he fell he struck his head against the kerbstone. He lay there, half-unconscious, but only half. The other half saw the advantage: if he lay there a little longer he might not have to go to school. From then on he had regular fainting fits, half real, half not, causing his parents so much worry that he was finally allowed to stay away from school for six months. 'A picnic,' he called it. But he also pitied his poor parents who were consulting many doctors, all in vain. No one could work out what was wrong with the boy. Finally it was decided he needed a change and he was sent off to stay with his architect uncle Ernst Jung in Winterthur. Carl loved it, spending hours at the town's railway station watching the steam trains come and go. But when he returned home to Klein-Hüningen he found his parents more worried than ever: he might have epilepsy, he overheard, and what were they to do, with no money and a boy who could not look after himself? 'I was thunderstruck. This was the collision with reality.' That same day he went to his father's library and started cramming. He had only one more fainting fit after that but did not let it master him, and soon he was back at school. 'That was when I learned what a neurosis is.'

From then on he got up at five every morning to study before setting off for school at seven. Sometimes it was 3 a.m. He felt he was himself for the first time. 'Previously I had existed too, but everything had merely happened to me. Now I happened to myself. Now I knew: I am myself now, now I exist.' In this elevated state he went to stay with a school friend who had a house on Lake Lucerne. How lucky the boy was, thought Carl, and how lucky they were to be allowed to use the *Waidling*, the punt, plunging the pole into the water as they manoeuvred out of the boathouse and into the blue. But when Carl started doing some fancy tricks, showing off, the boy's father whistled them back to shore and gave Carl a dressing down. Carl was seized with rage 'that this fat, ignorant boor should dare to insult ME'. But just as quickly he realised it was another conflict with reality: the father was right, he was wrong. It occurred to him that he might be two different people, the unsure boy and the 'other', the sure and powerful one. Not only did this other Carl exist, he was an old man, wore buckled shoes and a white wig and drove about in a fly with high wheels and a box suspended on springs with leather straps: a man living in the eighteenth century. The one as real as the other.

Around this time Carl was giving much thought to the idea of God. Not necessarily the God of his father's Protestant Reformed Church, but 'God the Creator', 'God of all Things'. One summer's day he came out of school and was again standing in the precincts of the cathedral – blue sky, radiant sunshine, gazing in awe at the pitched roof which had recently been retiled and glittered in the bright light – and thinking 'the world is beautiful and the church is beautiful, and God made all this and sits above it far away in the blue sky on a golden throne and . . . Here came a great hole in my thoughts and a choking sensation. I felt numbed, and only knew: Don't go on thinking now! Something terrible is happening, something I do not want to think, something I dare not even approach. Why not? Because I would be committing the most frightful of sins . . .' And on it went, all the way on the long walk home, all through that night, and the next. By the third night the feeling had become unbearable. 'Now it is coming, now it's serious,' he thought. '*I must think.*' The thinking brought him to the idea that it was God, the creator of this beautiful world, who wanted him to think, and, what's more, to think of something inconceivably wicked. In a way it had very little to do with him, he had no choice. Adam and Eve had been perfect creatures before they sinned, '*Therefore it was God's intention that they should sin.*'

That thought liberated him and he gathered all his courage to think about the cathedral, the clear sky, and God sitting high above it on His golden throne, 'and from under the throne an enormous turd falls upon the sparkling new roof, shatters it, and breaks the walls of the Cathedral asunder'. To his own amazement he felt an indescribable relief; and instead of damnation he felt grace had come down on him 'and with it an unutterable bliss such as I had never known'.

He never spoke about this to anyone, or about the other two secrets of the phallus dream and the manikin, until, finally, many years later, he told Emma. 'My entire youth can be understood in terms of this secret. It induced in me an almost unbearable loneliness. My one great achievement during those years was that I resisted the temptation to talk about it with anyone.'

His mother reminded him he had often been depressed as a boy, but by his mid-teens Carl's depressions gradually lifted. He read voraciously: Plato, Socrates, Hegel, Schopenhauer, Kant, Goethe, all the writers he would later introduce to Emma. His school friends started calling him 'Father Abraham'. Personality No. 1 was to the fore and he lived more in the present, active at school and out and about in Basel. But Personality No. 2 was never far away. One day, walking along the banks of the Rhine on his way home from school, he saw a sailing ship. A storm was blowing up and the mainsail was running before it. The sight propelled him into a detailed fantasy which would stay with him for the rest of his life: the river became a great lake with a high rock rising out of it, only connected to the mainland by a narrow causeway. A wooden bridge led to a gate flanked by towers opening into a little medieval town, and on the rock stood a castle: 'This was my house.' The rooms were panelled and simple, a fine library held everything worth knowing, and there were weapons and canons for protection as well as a garrison of fifty men at arms. The little town had several hundred inhabitants and Carl was the mayor, the justice of the peace, and general adviser. There was a small port on the landward side of the town where he kept his two-masted schooner. 'The raison d'être of this whole arrangement was the secret of the keep, which I alone knew. The thought had come to me like a shock. For, inside the tower, extending from the battlements to the vaulted cellar, was a copper column.' It was as thick as a man's arm and it stood like a tree, upside down with rootlets reaching into the air. These roots drew something from the air and conducted it down the column into the cellar where there was a laboratory in which Carl made gold out of the mysterious substance.

From this point on, Carl's long, boring walk home from school became a short, delightful one, lost in the fantasy, making structural alterations to the buildings, holding council sessions, sentencing evil-doers, firing canons. Alternatively he might go for a sail in his schooner, and before he knew it he was standing on the parsonage doorstep. The fantasy lasted several months before he bored of it. Thereafter he started building with mud and stones in the garden of the parsonage, he studied fortifications, collected fossils, learnt about plants and read numerous scientific periodicals. Building miniature castles and towns is what he used to do as a young boy to bring him back to himself, and he would do it for the rest of his life.

Carl passed his matriculation examinations with ease and went on to study medicine at the University of Basel. Originally he, like Emma, wanted to study the natural sciences, but he knew he needed to earn a proper living, and, deep down, he knew where he was going. After all, six of his mother's relations were pastors, healers of the spirit, and his Grandfather Jung, arriving in Basel from Germany, was a doctor of medicine with progressive views, believing that the insane should be given treatment, not incarcerated. Grossvater Jung was a well-known and respected figure around Basel – the kind called 'larger than life', a democrat and a liberal, somewhat eccentric, with a pink pig for a pet – in a word, just the kind of man Carl would wish to emulate. He never knew his grandfather but he shared his name: Carl Gustav, except his own spelling was with a K. He changed it to a C once he left university, embarking on his own life.

If Grossvater Jung was a liberal in public, at home he was authoritarian. His son, Paul Achilles Jung, Carl's father, found he could never live up to his expectations. Though Paul was a fine scholar, studying oriental languages and Hebrew at Göttingen, writing his dissertation on the Arabic version of the 'Song of Songs', when it came to choosing a profession he decided to become a pastor in the Protestant Reformed Church, a modest, retiring life, with a poor living. Perhaps he was encouraged in this by his future father-in-law, Samuel Preiswerk. Paul was Preiswerk's student and they spent many happy hours in his library going over ancient Hebrew texts. But something caused Paul Jung to always be racked with doubts. Most summers he went, alone, to stay with a Catholic priest in Sachseln. It was odd behaviour for a pastor of the Protestant Reformed Church, and no one knew why he went. In a way, Carl respected him for it.

But his respect for his father was ebbing away. By the time of his confirmation Carl was completely alienated from the

Church, bored and sceptical, arguing vehemently with his father about the hypocrisy of it all, and, worse still, his father's own hypocrisy. He watched his father go through the motions day after day, knowing what doubts and torments he suffered privately in the dark hours of the night. 'I was seized with the most vehement pity for my father,' he remembered in old age. 'An abyss had opened between him and me, and I saw no possibility of ever bridging it, for it was infinite in extent.' He was still his 'dear and generous father', but Carl could do nothing for him. He searched the Bible for answers but found none. Surely the devil was God's creature too? But of that the Bible gave no sign. It was nothing but 'fancy drivel'. In a letter of 13 June 1955, Jung admitted the tragedy of his youth had been to see his father 'cracking up' before his eyes.

Carl's mother, Emilie, came from a long line of seers. Her own mother had two personalities: a good monk and a bad monk, and she had visions and saw ghosts. The occult was part of everyday life for many of the Preiswerks and once Carl's mother came back home after being 'away', the occult became part of his everyday life too, or more often his night life when alarming 'atmospheres' emanated from her bedroom. 'I was sure she consisted of two personalities, one innocuous and human, the other uncanny,' he wrote. In her Personality No. 1 she was a good mother: warm, pleasant, with a good sense of humour, and a fine cook, and inclined to look up to her son as he grew older, confiding in him instead of in his father. But her Personality No. 2 was a different matter: 'a sombre, imposing figure possessed of unassailable authority – and no bones about it'. She was also very large and overweight. Later he realised this was due to her depression and the bad state of the marriage, but as a boy he did not understand, and his deep, complicated ties were to his clever, tormented father. 'The feeling I associated with "woman" was for a long time that of innate unreliability. "Father", on the other hand, meant reliability and – powerlessness. This is the handicap I started off with. Later these early impressions were revised: I have trusted men friends and been disappointed by them, and I have mistrusted women and was not disappointed.'

By the time Carl enrolled in Basel University in early 1895 his father had become, in his eyes, like the Fisher King from the Grail legend 'whose wound would not heal', and had begun to show signs of a real wound which would not heal. He had been depressed and tormented for years but now the family doctor found serious physical symptoms which he could not diagnose, but which were nevertheless killing him. By the end of the year he was bedridden. Carl carried him from room to room like a bag of bones. Within months he was dead. 'The following days were gloomy and painful,' wrote Carl. 'Little of them has remained in my memory.'

Before he died Carl's father had applied to the canton of Basel for a stipend to help fund his son's studies. The request was granted, but instead of being pleased Carl was mortified at having to resort to charity. He also had to borrow money from his wealthy Jung relations and he bought and sold antiques for one of his aunts to make ends meet. As to his mother, Emilie: 'Once my mother spoke to me or to the surrounding air in her "second" voice, and remarked: "He died just in time for you."'

At the time Carl was not sure what she meant, but his father's death certainly freed Carl: his student years were a good time, full of energy, friendship and intellectual activity, fuelled, as he himself recognised, by a 'vaulting ambition'. Soon he was dominating the discussions at his Zofingia fraternity, challenging the others with brilliant forays into the ideas of Schopenhauer and Nietzsche and talking about things unknown: dreams, visions, and the occult. His friend, Albert Oeri, progressing with him from the *Gymnasium* to Basel University, remembered discussions on subjects such as: 'The Limits of Exact Science' and 'Some Reflections on the Nature and Value of Speculative Research' and 'Some Thoughts on Psychology', and that Carl easily succeeded in 'intellectually dominating an unruly chorus of fifty or sixty students from different branches of learning, and luring them into highly speculative branches of thought'. He quickly acquired the nickname Walze – 'the Steam-Roller' – and looking at fraternity photographs of Carl at the time, anyone can see why: large and round, he has an overbearing presence and a less than charming, closed expression. His vaulting ambition and superior manner did not endear him to everyone. As to women, there weren't any.

When his father died, Carl and his mother and his sister Trudi, nine years his junior, had to leave the parsonage. They had no money and nowhere to go so they moved in with the Preiswerk family in an old, dilapidated mill, the Bottminger Mill, in a run-down district on the outskirts of Basel. Moving into the mill brought Carl into direct contact with spiritualism and the occult because this was the branch of the family who were seers, had visions, heard voices and held seances. 'He was appalled that the official scientific position of the day towards occult phenomena was simply to deny their existence,' wrote Oeri, 'rather than investigate and explain them.' Consequently

Carl decided to investigate them, regularly attending seances at the mill led by his cousin Helly, and subsequently basing his doctoral dissertation on them, entitled 'On the Psychology and Pathology of So-Called Occult Phenomena'.

So this was the man Emma met, or became reacquainted with, in 1899 on her return from Paris, aged seventeen. He was twenty-four, just starting work at the Burghölzli asylum, a complex man with many secrets. In fact there was one further secret Emma did not know about the 'other' Carl. It was one he himself no longer 'knew', having repressed it deep in his unconscious where it safely remained until that wizard of the unconscious, Sigmund Freud, uncovered it during one of those long, intense evenings the two men shared in March 1907.

It threw Carl right off balance, as Emma cannot have failed to notice. Especially in the way he behaved with the Jewish woman they met in the hotel in Abbazia. He did not admit it to Emma then, or for many years. At the time he did not even admit it to himself – not until the crisis had become so unmanageable that he could no longer avoid it: Carl had been sexually abused when he was a boy and the only way he could deal with it was to repress the memory. But in October 1907, a good seven months after the discussions with Freud had forced the memories to the surface, and after his life had been thrown into further confusion, he had to confront it. His answers to Freud's letters had been more and more delayed, and finally Freud, usually so tolerant and indulgent of his crown prince, voiced his objections and Carl came clean. 'Your last two letters contain references to my laziness in writing. I certainly owe you an explanation,' he wrote on 28 October, first blaming his workload but then admitting that it was actually what Freud had termed his 'self-preservation complex' which often bedevilled his pen, preventing him from writing:

> Actually – and I confess this to you with a struggle – I have a boundless admiration for you both as a man and a researcher, and I bear you no conscious grudge. So the self-preservation complex does not come from there; it is rather that my veneration for you has something of the character of a 'religious' crush. Though it does not really bother me, I still feel it is disgusting and ridiculous because of its undeniable erotic undertone. This abominable feeling comes from the fact that as a boy I was the victim of a sexual assault by a man I once worshipped. Even in Vienna the remarks of the ladies (*'enfin seuls'* etc.) sickened me, although the reason for it was not clear to me at the time. This feeling, which I have still not quite got rid of, hampers me considerably.

He goes on to explain that this has made close friendships with male colleagues 'downright disgusting', and ends the letter abruptly, saying: 'I think I owe you this explanation. I would rather not have said it.'

All these secrets, and Emma knew none of them.

3

A Secret Betrothal

Emma and Carl were betrothed on 6 October 1901, in secret, at the family's Ölberg estate in Schaffhausen. The only other guests apart from Emma's mother and sister were Carl's mother and sister and the only record of the event are a few out-of-focus snapshots, probably taken by Marguerite, in the garden, where the couple appear to be walking away from the camera as often as towards it, as though trying to avoid it. By the time of the betrothal they had been courting for eighteen months and Emma was nineteen. She looks younger, almost a schoolgirl, and Carl still has the look of the Steam-Roller about him, not yet the confident man of the world he would become. They agreed to wait until Emma was twenty-one before getting married.

Jung had been assistant physician at the Burghölzli asylum for almost a year by then, having moved to Zürich straight after his medical examinations, taking a temporary job in a doctor's practice to fill in time before starting at the Burghölzli in December 1900. The move puzzled his colleagues and upset his mother. Why would anyone want to leave cosmopolitan Basel for dull, commercial Zürich, and how could he abandon his impoverished mother and sister like that? Part of the reason was that Carl had fallen foul of Herr Professor Wille, the first in a long line of senior men Jung would quarrel with in his working life. Ludwig Wille was the new professor of psychiatry at the University of Basel, the discipline of psychiatry itself dating no further back than the 1880s. In deciding to specialise in it, Jung embarked on the least fashionable and least remunerated branch of the medical profession, seen as another odd decision by his colleagues, given he was one of their best with the brightest of futures. They could not know that psychiatry spoke deeply to the 'other' Carl who had had, ever since he could remember, the kind of dreams and visions some people might deem insane. When he read Krafft-Ebing's 1890 work on 'diseases of the personality' in the *Lehrbuch der Psychiatrie*, he recognised it immediately. 'My heart suddenly began to pound. I had to stand up and draw a deep breath. My excitement was intense . . . Here was the empirical field common to biological and spiritual facts, which I had everywhere sought and nowhere found.' Professor Wille, however, was firmly of the old school, seeing all mental illness as the result of a physical deterioration of the brain. Sooner or later Jung was bound to take issue with that, the net result being that he decided to present his doctoral dissertation to the medical faculty of the University of Zürich, not Basel. The city had the added advantage of being closer than Basel to Schaffhausen and Emma.

There were around 400 patients at the Burghölzli asylum when Jung arrived to take up his position, but apart from Bleuler there was only one other qualified doctor, Ludwig von Muralt, the day-to-day care being carried out by unqualified helpers, male and female. This meant an extremely heavy workload for Carl, who also lived on the premises. The only time he had off was Sundays, when he was able to visit Emma, first walking fifteen minutes down the hill to the tram station which in those days did not reach as far as the Burghölzli, then taking the recently electrified tram to Zürich's Central Station, then the steam train to Schaffhausen, speeding northwards leaving lake and alps behind and into a landscape of farmland, meadows and scattered villages, past the crashing Rhine Falls, finally pulling in to Schaffhausen railway station where he was met by coachman Braun in the green Rauschenbach carriage and conveyed up the hill to the Ölberg estate – thereby travelling from the lowest to the highest *niveau* of Swiss society all in the course of two hours, a journey which never failed to delight.

And there in the grand front hall waiting expectantly for him was Emma, eager to listen, eager to learn, eager to help. She had started taking Latin lessons in order to read Carl's medical texts, and maths lessons to discipline her mind, and she was practising her handwriting to help Carl with the daily reports which Bleuler was most particular about, regardless of how many hours it took to write them up. Boring for Carl, but thrilling for Emma. As to her general knowledge: as with Carl, the natural sciences were her enthusiasm; so too her interest in the Legend of the Holy Grail, which had likewise been Carl's for many years. Emma meant to help Carl with his work, if only in small ways, and Carl was only too happy to oblige.

In many ways Emma was preparing to be 'the good wife' in similar vein to those recommended to young ladies by the weekly magazines. 'The ideal wife,' wrote Rudolf von Tavel in the monthly *Wissen und Leben*, 'should live and act entirely in her husband's spirit. She must support him in his task, softening him, warming him, and praising him in golden terms, convincing her children of the same, so that the way of life in the family is the right one, fostering the right social attitudes for the upholding of the Vaterland.' Rosa Dahinden-Pfyl agreed, writing in *Die Kunst mit Männern Glücklich zu Sein* (*The Art of Being Happy with a Man*): 'The happiness and lasting power of married love relies largely on the good and clever ways of the woman.' She should never complain about the husband's coldness towards her, whether real or imagined. She should take a gentle interest in everything which concerns him, always showing her appreciation, and make his home comfortable, never tiring him with needless chatter. She should avoid becoming bitter about his weaknesses, and never meddle in his business affairs. To remain attractive to him she should always be sweet-tempered, dress nicely and with good taste; in fact, always take care of her appearance but also her health, her character and her soul. 'Should her physical charms fade, she should retain her husband's interest by her sympathy, her learning, her heart and spirit, but never by showing a knowledge greater than his.'

Had Emma married her haut-bourgeois beau this would surely have been her whole life, and nothing more. But not with Carl. His own childhood had nothing bourgeois about it and his own character was not suited to fitting in with convention. His requirements of Emma would be infinitely more complex and challenging than any handbook on marriage could encompass. But all that came later. For the time being it was simply exhilarating for Emma to be with someone who brought her books and scientific articles to read and was happy to discuss his plans and ideas with her.

Carl and Emma easily fell into the roles of teacher and student. Carl had already completed five years of medical studies and embarked on the work and profession that would occupy him for the rest of his life. At this time he was working on his doctoral dissertation, 'On the Psychology and Pathology of So-called Occult Phenomena', investigating the uncharted world of the unconscious through the evidence he had gathered during Helly's seances at Bottminger Mill. Herr Professor Freud in Vienna had just published a book called *The Interpretation of Dreams*, which investigated the unconscious from a different angle, the hidden meaning of dreams, which had often filled the pages of literature but had never yet been scientifically investigated, and Carl had been deputed by Bleuler to read it and present his findings to the Burghölzli staff at one of their evening meetings. As if he didn't already have enough paperwork to do. But this was different. This was the new world, just waiting to be discovered.

To Emma, Carl was worldly and sophisticated and, as his friend Albert Oeri described, he was a mesmerising talker. So Emma might have been surprised to discover that her fiancé had absolutely no experience of women. 'He didn't think much of fraternity dances, romancing the housemaids, and similar gallantries,' recalled Oeri, making Carl sound like a bit of a prig. He did bring Luggi, Helly's older sister, to some dances, but Oeri only remembered one occasion when Carl was really smitten, at another fraternity dance, with a young woman from French-speaking Switzerland, at which point he began to behave very oddly indeed. In fact, if Oeri had not been such a reliable witness, one would wonder at the veracity of the story. 'One morning soon after,' Oeri wrote, 'he entered a shop, asked for and received two wedding rings, put twenty centimes on the counter, and started for the door.' Presumably the twenty centimes were by way of a deposit. When the owner objected,

Carl gave back the rings, took his money and left the shop 'cursing the owner, who, just because Carl happened to possess absolutely nothing but twenty centimes, dared to interfere with his engagement'.

With anyone else you might think this was some kind of student prank, but not with Carl, hovering precariously between two personalities and having no idea how to handle such a situation. Afterwards 'Carl was very depressed,' said Oeri. 'He never tackled the matter again, and so the Steam-Roller remained unaffianced for quite a number of years.' Until he met Emma, in fact, and persuaded her to marry him.

But Emma had a secret of her own. When she was twelve her father started to lose his sight. Soon he could no longer read for himself, and Emma, the studious one, was deputed to sit and read out loud to him: newspapers, magazines, books, business matters brought over from the factory and foundry nearby, so she became quite knowledgeable in financial matters and the handling of accounts. Later, as his condition worsened, plans for the redevelopment of the Ölberg estate were made for him in Braille. It was hard for Emma, not least because her father was a difficult, sarcastic man – a trait which got worse with age and the advance of his illness. Harder still was the fact that the cause of his blindness had to be kept secret, such was the shame and stigma attached to it: Herr Rauschenbach had syphilis. According to the family, he had caught the disease after a business trip to Budapest, presumably from a prostitute. Bertha had decided not to go with him on that occasion because she felt the two little girls were too young to be left alone with the children's maid. Had she gone, everything might have been different. It was a tragedy for the family, and photographs of

Emma during her teens show a shy, round-faced, podgy girl, surely feeling the stress of the family secret. She was trying her best to help her mother with this awful burden as her father became more and more bitter and desperate, shut away from the world in his room upstairs. It robbed Emma of her sunny nature, making her too serious for her age.

Albert Oeri remembered visiting the Jung household not long before Pastor Jung died and described how Carl, aged twenty, carried his father 'who had once been so strong and erect' around from room to room 'like a heap of bones in an anatomy class'. Emma sat at her father's side reading to him as he went blind, bitter and half mad. There was not much to choose between them.

By the end of the nineteenth century doctors were finally on the verge of finding a cure for syphilis, but not soon enough for Herr Rauschenbach. By 1905 Fritz Schaudinn and Erich Hoffmann in Berlin had identified the causative organism, the microbe *Treponema pallidum*, and by 1910 Dr Paul Ehrlich, director of the Royal Prussian Institute of Experimental Therapy in Frankfurt, developed the first modestly effective treatment, Salvarsan, though it was not until the discovery of penicillin in the 1940s that a cure was certain. In the 1890s the treatment still relied on the use of mercury, which could alleviate the condition if caught early enough though the side effects were extremely unpleasant, and it was not a cure. Syphilis, highly infectious and primarily transmitted through sexual contact, had stalked Europe for centuries, causing fear and dread and giving rise to a great deal of moralising about the virtues of marriage. The symptoms were horrible, the first signs being rashes and pustules over the body and face, then open suppurating lesions in the skin, disfiguring tumours and terrible pain, only alleviated by regular doses of morphine. Some of the most tragic cases were those of unsuspecting wives infected by their husbands, in turn infecting the unborn child. Wet nurses were vulnerable, either catching it from the child, or, already infected themselves, passing it on to the child instead. To make matters worse for families like the Rauschenbachs, society was hypocritical about syphilis. Everyone knew about the disease but it was not talked about, except in the medical pamphlets read in the privacy of a doctor's surgery: 'The woman must submit to her husband – consequently, whereas he catches it when he wants, she also catches it when he wants! The woman is ignorant . . . particularly in matters of this sort. So she is generally unaware of where and how she might catch it, and when she has caught it she is for a long time unaware of what she has got.' Another pamphlet concentrated on women of the lower classes, unwittingly revealing a further hypocrisy of the times: whilst the bourgeois woman was seen as the victim, the working-class woman, not her seducer, or client if she was a prostitute, was to blame: 'The woman must be told . . . Every factory girl, peasant and maid must be told that if she abandons herself to the seducer then not only does she run the risk of having to bring up the child which might result from her transgression, but also that of catching the disease whose consequences can make her suffer for the rest of her life.'

In Vienna, Sigmund Freud was investigating the psychological effects of syphilis on the next generation, finding that many of his cases of hysteria and obsessional neurosis, such as his patients 'Dora' and 'Rat Man', had fathers who had been treated for syphilis in their youth. It is possible that Bertha

Rauschenbach was keen on Carl Jung as a suitor for Emma not only because she could see how well suited they were, but because Carl had just completed his medical studies and could help with the treatment of her husband's illness, in secret, in the privacy of their own home. And she was no doubt relieved to realise how little experience her future son-in-law had had with women.

Meanwhile Carl was working at the Burghölzli, putting his father, who had died 'just in time' as his mother said, behind him. But he could not leave Personality No. 2 behind. Years later his friend and colleague Ludwig von Muralt, the other doctor at the Burghölzli when Carl arrived, told him that the way he behaved during those first months was so odd people thought he might be 'psychologically abnormal'. Jung himself described experiencing feelings of such inferiority and tension at the time that it was only by 'the utmost concentration on the essential' that he managed not to 'explode'. The problem was partly that Bleuler and Von Muralt seemed to be so confident in their roles, whereas he was completely at sea in this strange new world of the institution, and partly because he felt deeply humiliated by his poverty. He had only one pair of trousers and two shirts to his name and he had to send all his meagre wages back to his mother and sister, still living on charity at the Bottminger Mill. The humiliation was accompanied by a general feeling of social inferiority, heightened by the fact that Von Muralt came from one of the oldest and wealthiest families of Zürich.

They were the same feelings which had often plagued Carl in the past and his solution was the same: to withdraw into himself and become what he called a 'hermit', locking himself away from the world. When he was not working, he read all fifty volumes of the journal *Allgemeine Zeitschrift für Psychiatrie*, cover to cover. He said he wanted to know 'how the human mind reacted to the sight of its own destruction'. He might have used his own father as an illustration. Or himself during those first months of 1901 after Emma had refused his hand in marriage, perhaps the real reason why he was so distressed, when his confident Personality No. 1 disappeared into thin air along with all his hopes and dreams, and he was close to a mental and emotional breakdown, as he had been in the past and would be again in the future.

The crisis was extreme. But then, quite suddenly, at the end of six months, he recovered. In fact, he swung completely the other way, the inferior wretch replaced almost overnight by the loud, opinionated, energetic Steam-Roller of old. Once betrothed and sure of Emma's love, Carl was able to take life at the Burghölzli at full tilt, with the kind of energy which left others breathless. No one could fail to notice it, but no one knew the reason why, because no one knew where he went every Sunday on his day off.

Not knowing of Carl's extreme crisis of confidence, Emma remained dazzled by his love, hardly able to believe it was true. She worried that she was a boring companion as she recounted small details about her mundane week – the riding, the walks by the Rhine, the family visits, her father's deteriorating health, the musical evenings which her mother liked to host at Ölberg. The best moments were when they discussed the books she'd been reading, which gave her week its shape and purpose, enabling her to be 'herself' in a way which would otherwise have eluded her. To her joy Carl was delighted by her progress, always encouraging her to do more. And to her relief she soon discovered he was not in the least interested in having a bourgeois wife who thought of nothing but home and children and life within the narrow confines of Swiss society. Every day she waited for the postman, struggling up the hill to Ölberg on his bicycle, bearing another letter from Carl addressed to '*Mein liebster Schatz!*' – my darling treasure – long letters, filled with Burghölzli news, ideas and suggestions for further reading, and telling her how much he loved her. And every Sunday, as Carl got to know Emma better, he found that beneath the shyness and seriousness there hid another Emma: one with a lively sense of humour, who could laugh and laugh. And who better to make her laugh than Carl?

The Burghölzli at the turn of the twentieth century under the directorship of Bleuler was a remarkable institution rapidly gaining an international reputation. At a time when most asylums simply removed the insane from society, locking them up, often for whole lifetimes, the Burghölzli offered treatment of various kinds and tried, as far as possible, to show the patients consideration and respect. Jung himself describes the situation:

> In the medical world at the time psychiatry was quite gener-
> ally held in contempt. No one really knew anything about it,
> and there was no psychology which regarded man as a whole
> and included his pathological variations in the total picture.
> The Director was locked up in the same institution with his
> patients, and the institution was equally cut off, isolated on
> the outskirts of the city like an ancient lazaretto with its
> lepers. No one liked looking in that direction. The doctors

knew almost as little as the layman and therefore shared his feelings. Mental disease was a hopeless and fatal affair which cast its shadow over psychiatry as well . . .

Soon after Jung joined the staff numbers increased to five doctors, and as far as Bleuler was concerned five was a luxury; before he took the job of director of the Burghölzli in 1898 he had spent thirteen years as director of the lunatic asylum on the island of Reichenau, where there were over 500 inmates with only one trained medical assistant.

Eugen Bleuler was a remarkable man. Coming from Swiss peasant stock, he was the first of his family to attend university, and, much like Jung, he was drawn to this new branch of medicine because he had experience of mental illness in his own family. His sister, Pauline, was a catatonic schizophrenic and after Bleuler married in 1901 she lived with him, his wife and their eventual five children in a large apartment on the first floor of the Burghölzli. Bleuler had trained with some of the most progressive practitioners of the age, including Dr Jean-Martin Charcot at the Salpêtrière Hospital in Paris, where Sigmund Freud also spent time. Charcot was first and foremost a neurologist concerned with the functions and malfunctions of the brain, demonstrated with a showman's flair to the hundreds of students who flocked to his lectures from England, Germany, Austria and America. He followed his patients' progress throughout their lives and when they died he examined their brains under the microscope, making early diagnoses of Parkinson's, multiple sclerosis, motor neurone disease and Tourette's. But it was his patients suffering from hysteria who interested him most and brought him his greatest fame. The fashionable 'treatments' at the time for hysteria were 'animal magnetism' and hypnotism, each requiring a doctor with special 'intuition'. In animal magnetism the doctor passed his hands over the patient to release the vital fluid or energy which had supposedly become blocked. In hypnotism the doctor took control of the mind, providing the most dramatic demonstrations as patients fell into trances and spoke in strange voices. When Eugen Bleuler continued his training under Auguste Forel at the Burghölzli in the early 1880s, hypnotism was one of the main treatments and Forel had his own *'Hypnosestab'*, a wand with a silver tip which he used with some success on obsessives and neurotics as well as hysterics. But his greatest success was with alcoholics. The Burghölzli was then, and remained when Bleuler took over from Forel, an institution which held strictly to the virtues of abstinence. Drink was one of the worst afflictions of the age and asylums were full of chronic cases who were contained whilst 'inside', but, without a vow of abstinence, soon reverted to old habits once they left.

All this Carl explained to Emma in his letters, or on their Sundays in the drawing room at Ölberg, or on their afternoon walks high up in the meadows above the house, up to 'their' bench by the edge of the forest beyond. Later Emma confessed she only understood half of it at the time. Apparently Bleuler was continuing the progressive methods he had started to develop at Reichenau: staff lived amongst the inmates, eating with them at the same tables and socialising with them in their spare time. His theory of *affektiver Rapport*, listening with empathy, was the guiding principle. There was also a great emphasis put on cleanliness. Inmates were helped to wash thoroughly, in spite of a shortage of baths and bathrooms, and to keep their clothes in good order; likewise their beds, a dozen on each side of a ward, which were kept neat, the heavy feather covers hung out of the windows every morning to air and the mattresses regularly turned. The patients were kept occupied, Herr Direktor Bleuler believing that physical activity was good for the distraught mind. The kitchen gardens lay beyond the walls of the asylum extending down the slopes and provided all their vegetables and fruit, which were brought fresh to the kitchens every day. The dairy produced the milk and cheese. Hens provided the eggs. The laundry kept inmates busy washing and starching and ironing. The *Hausordnung* kept the building spick and span, smelling of floor polish and soap. There were workshops: woodwork and wood-chopping for the tiled stoves in winter, sack-making, silk-plucking, sewing, mending, knitting. The place was to all intents and purposes self-sufficient and the food, for an institution, was good: always a soup for the midday meal followed by meat and vegetables, with soup and bread again for the evening meal along with a piece of cheese or sausage. The patients were divided into three categories and while third-class patients did not eat as well as the first-class (private) ones, it was still a better standard than at most asylums. In the evenings there were card games, reading, concerts; sometimes the patients produced an entertainment, sometimes lectures were put on, and at the weekends there were occasional fetes and dances for those willing and able. Jung was put in charge of social events as soon as he arrived. He hated it.

To keep things going day to day there were seventy *Wärters*, male and female helpers in long white aprons and starched white collars, the men in trousers, shirt and tie, the women in long dark skirts and starched white caps. Like the doctors they lived on the premises, but unlike the doctors they had no quarters of their own. They slept on the wards or in the corridors, on wooden camp beds put up for the night, the women in the women's section, the men in the men's. The only exception was the *Wärters* in charge of the 'first-class' patients, who would sleep in the patient's private room. This was a great privilege because not only was there some peace and quiet but the private patient was allowed candles in the room at night, or even an oil lamp if their behaviour was good enough and they were no danger to themselves or others. It was an eighty-hour week and the pay was low: 600 Swiss francs per annum for the male *Wärters*, a hundred francs less for the women. But board and lodging was all found, so the rest could be saved.

Carl's day started at 6 a.m. and rarely finished before 8 p.m., after which he would go up to his room to write his daily reports, work on his dissertation, and compose his daily letter to Emma. There was a staff meeting every morning, after a breakfast of bread and a bowl of coffee, ward rounds morning and evening, an additional general meeting three times a week to consider every aspect of the running of the institution, and at least once a week a discussion evening, which, by the time Carl joined, was already well versed in the writings of Freud. Once into his stride there was no stopping Herr Doktor Jung, and his sheer brilliance soon singled him out. Bleuler's *affektiver Rapport* was exactly the kind of treatment he himself believed in: listening acutely and with empathy to the apparent babblings of inmates with *dementia praecox*, or the outbursts of hysterics and the circular repetitions of obsessive neurotics. Carl was fascinated by the chronic catatonics who had been at the Burghölzli for as long as anyone could remember, including the old women incarcerated since they were young girls for having illegitimate children, who no longer knew who they'd once been.

Carl could listen to his patients for hours, taking notes, catching clues, watching how the mind worked. The Burghölzli was known for its progressive research, and inmates demonstrating interesting symptoms were the willing, and unwilling, guinea pigs – ushered into the room or lecture theatre for the doctor to examine, question and offer a diagnosis. It was a fine apprenticeship for Carl, as he himself admitted. He had little interest in the patients who were there with TB or typhus, and he could not bear the routine work, the meetings and the administration, to which he rarely gave any proper attention. But the old lady who stood by the window all day waiting for her long-lost lover, or the schizophrenic who talked crazily about God – that was a different matter altogether. Here his Personality No. 2 came into its own, working hand in hand with No. 1. 'It was as though two rivers had united and in one grand torrent were bearing me inexorably towards distant goals,' he later wrote. Bleuler soon saw that Herr Doktor Jung, with his fine intuition on the one hand and his brilliant mind on the other, understood the inmates like no one else.

Carl's listening was made more effective by Bleuler's insistence that everything be conducted in Swiss dialect, not the High German which doctors normally used, and which effectively meant no communication since most inmates could not understand High German, and even if they could, it caused such a social gulf between doctor and patient that the patient felt browbeaten. When speaking High German, Carl retained his broad Basel accent, something which had humiliated him at grammar school, but no longer. Now he learnt all the other regional Swiss dialects as well because at the Burghölzli it was expected that the doctor should adapt to the patient, not the other way round, an idea generally held as odd and even danger-ous by the vast majority of the medical profession. Besides, without it Jung could do no useful research.

On Sundays he retold the patients' stories to Emma, shock-ing her, entertaining her, keeping her spellbound. Stories about women in the asylum held a special fascination for her, such as the woman in Carl's section who had been diagnosed with schizophrenia. Jung disagreed with the diagnosis, thinking it was more like ordinary depression, so he started, *à la Freud*, to ask the woman about her dreams, probing her unconscious. It turned out that when she was a young girl she had fallen in love with 'Mr X', the son of a wealthy industrialist. She hoped they would marry, but he did not appear to care for her, so in time she wed someone else and had two children. Five years later a friend visited and told her that her marriage had come as quite a shock to Mr X. 'That was the moment!' as Jung wrote in his account of it. The woman became deeply depressed and one day, bathing her two small children, she let them drink the contaminated river water. Spring water was used only for drinking, not bathing, in those days. Shortly afterwards her little girl came down with typhoid fever and died. The woman's depression became acute and finally she was sent to the Burghölzli asylum. Jung knew the narcotics for her *dementia praecox* were doing her no good. Should he tell her the truth or not? He pondered for days, worried that it might tip her further into madness. Then he made his decision: to confront her, tell-ing no one else. 'To accuse a person point-blank of murder is no small matter,' he wrote later. 'And it was tragic for the patient to have to listen to it and accept it. But the result was that in two weeks it proved possible to discharge her, and she was never again institutionalised.'

Love, murder, guilt, madness. Emma had never heard stories like it. Nor had she ever considered the powerful workings of the 'unconscious'. But she was intelligent and well read. She knew all about Faust's pact with Mephistopheles, Lady Macbeth's guilty sleepwalking, and Siegfried, the mythical hero. Now she was beginning to realise that this 'unconscious' was the key to the hidden workings of the mind and could be accessed in various ways, then used as a tool to cure the patient. By the time of their secret betrothal Emma was helping to write up Carl's daily reports, learning all the time. If these years were the beginning of Carl Jung's career, they were the beginning of something for Emma too.

Meanwhile Jung was trying, between his eighty-hour week and his social duties, to finish his dissertation on the 'So-called Occult Phenomena'. There had been a revival of interest in the occult at the end of the nineteenth century, people using Ouija boards and horoscopes, having seances, delving into magic and the ancient arts, and reporting strange paranormal occurrences. Carl was used to such things from his mother's 'seer' side of the family, like the time when a knife in the drawer of their kitchen cupboard unaccountably split in two, or the times when his mother spoke with a strange, prophetic voice. But what interested Jung the doctor was the way the occult provided another route to the hidden world of the unconscious, more psychological than spiritual. Cousin Helly was probably a hysteric, he now concluded, a young girl falling into trances to get attention.

In fact Helly's mother had become so worried about the way the trances and the voices were dominating her life that she had packed her off to Marseilles to study dress-making, at which point Helly's trances stopped, and she became a fine seamstress.

Jung presented his dissertation to the faculty of medicine at the University of Zürich in 1901 and it was published the following year. The Preiswerk family, reading it on publication, were distressed. After a general introduction about current research on the subject, Jung concentrated on one case history: Helly Preiswerk and her seances at Bottminger Mill, referring to her, by way of thin disguise, as Miss S. W., a medium, fifteen and a half years old, Protestant – that is, instantly recognisable by anyone who knew the family. She was described as 'a girl with poor inheritance' and 'of mediocre intelligence, with no special gifts, neither musical nor fond of books'. She had a second personality called Ivenes. One sister was a hysteric, the other had 'nervous heart attacks'. The family were described as 'people with very limited interests' – and this of a family who had come to the rescue of his impoverished mother and sister. It showed a side of Carl which Emma would have to deal with often in the future: a callous insensitivity, driven perhaps by what he himself admitted was his 'vaulting ambition'. But as far as the faculty of medicine at the University of Zürich was concerned, it was a perfectly good piece of research, fulfilling the aims Jung expressed in his clever conclusion: that it would contribute to 'the progressive elucidation and assimilation of the as yet extremely controversial psychology of the unconscious'.

The most telling thing about the dissertation is the dedication on the title page. It reads: 'to his wife Emma Jung-Rauschenbach'.

Given the dissertation was completed in 1901 and published in 1902, the dedication precedes the event of the marriage by a good year. Whatever was Carl thinking? What is the difference between 'my wife', which Emma was not, and 'my betrothed', which she was. It suggests Carl was desperate to claim Emma for himself, fully and legally, a situation which mere betrothal could not achieve. Emma was the answer to all Carl's problems: financial – certainly – but equally his emotional and psychological ones. He needed Emma for his stability in every sense, and he knew it.

Carl and Emma's wedding was set for 14 February 1903, St Valentine's Day. Buoyed up and boisterous, Carl became increasingly impatient and intolerant of the 'unending desert of routine' at the Burghölzli. Now he saw it as 'a submission to the vow to believe only in what was probable, average, commonplace, barren of meaning, to renounce everything strange and significant, and reduce anything extraordinary to the banal'. Given Bleuler's achievements at the asylum this was high-handed Carl at his worst. The fact is, he was fed up. He wanted to take a sabbatical, to travel, to do things he had never been able to do before. And he wanted to do it before he married Emma. Because he had no money of his own, his future mother-in-law happily offered to fund it. In July 1902 Jung submitted his resignation to Bleuler and the Zürich authorities, and by the beginning of October he was off, first to Paris, then London, for a four-month pre-wedding jaunt. Bleuler, knowing nothing of Jung's secret betrothal to Fräulein Rauschenbach, must have been angry and non-plussed. How would Herr Doktor Jung afford it? How could he manage without a salary? And what about his poor mother and sister?

Emma and her mother meanwhile started the lengthy process of preparing for the wedding – the dress itself, the veil, shoes, bouquet, trousseau, church service, flowers, guest list, menu for the wedding banquet, and the travel arrangements for the couple's honeymoon. Emma's father's health must have caused some heartache: already parlous, it had taken a turn for the worse. He knew about his daughter's betrothal to Carl now, but what to do about the wedding? It was a terrible dilemma for Emma who had, over the years, watched her father's decline in horror and shame, and now he would not be able to attend the ceremony, or walk her up the aisle, or give her away.

Before leaving the country Carl had to complete his Swiss army military service, an annual duty for all Swiss males between the ages of twenty and fifty, in his case as a lieutenant in the medical corps. But then he was off, first paying a visit to his mother and sister in Basel on his way to France. From Paris he wrote daily letters to Emma, and separately to her mother too, giving them all the news: he lived cheaply in a hotel for one franc a day and worked in Pierre Janet's laboratory at the Salpêtrière, attending all the eminent psychologist's lectures. He had enrolled at the Berlitz School to improve his English and started reading English newspapers, a habit he retained all his life. He went to the Louvre most days, fell in love with Holbein, the Dutch Masters and the *Mona Lisa*, and spent hours watching the copyists make their living selling their work to tourists like himself. He walked everywhere, through Les Halles, the Jardin du Luxembourg and the Bois de Boulogne, sitting in cafés and bistros watching the world, rich and poor, go by, and in the evenings he read French and English novels, the classic ones, never the modern. He also saw Helly and her sister Vally, both now working as seamstresses for a Paris fashion house, and he was grateful for their company, not only because he knew no one else in Paris, but because Helly was generous enough to forgive him his past sins. When the weather turned cold Bertha Rauschenbach posted off a winter coat to keep him warm. It wasn't the only thing she sent him: when he expressed a longing to commission a copy of a Frans Hals painting of a mother and her children, the money was quickly dispatched.

By January Carl was in London, visiting the sights and the museums and taking more English lessons. It must have been his English tutor, recently down from Oxford, who delighted Carl by taking him back to dine at his college high table with the dons in their academic gowns – a fine dinner, as he wrote to Emma, followed by cigars, liqueurs and snuff. The conversation was 'in the style of the 18th century' – and men only, 'because we wanted to talk exclusively at an intellectual level'. In 1903 it was still common for men to be seen as more intellectual than women, and there were no women at high table to disagree. It was Jung's first brush with the English 'gentleman' and he never forgot it, the word often appearing in his letters as a mark of the highest praise.

In Schaffhausen, Emma received a present. It was a painting. In between all his other activities Carl took the time in Paris to travel out into the flat countryside with his easel and paints. He had always loved painting, even as a child, and in future it would go hand in hand with his writing, offering a poetic and spiritual dimension to his words. He found a spot on a far bank of the River Seine, looking across at a hamlet of pitched-roof houses, a church with a high spire, and trees all along. But the real subject of the painting was the clouds, which took up three-quarters of the canvas: light and shining below, dark and dramatic above. The inscription read: 'Seine landscape with clouds, for my dearest fiancée at Christmas, 1902. Paris, December 1902. Painted by C. G. Jung.'

It might have been a premonition of their marriage.

The painting Carl sent Emma for Christmas 1902.

4

A Rich Marriage

Emma and Carl got married twice: first the civil ceremony on 14 February, with just a few family members in attendance – Carl's mother and sister, and Emma's sister Marguerite but not her mother or father – followed by an evening wedding ball at the Hotel Bellevue Neuhausen overlooking the Rhine Falls. Then, two days later, a church wedding in the Steigkirche, the Protestant Reformed Church in Schaffhausen, at which both mothers were present, but not Emma's father. The bridal couple arrived at the church in the high Rauschenbach carriage used only for weddings, decorated with winter flowers and driven by coachman Braun: Carl in his newly acquired silk top hat and tails, Emma in her white gown, veil and fur cape against the cold – luckily there was no snow that February – waving and smiling at well-wishers left and right. The service was followed by a wedding banquet at Ölberg. Later, on 1 March, there was a further festivity at the Hotel Schiff for the employees of the Rauschenbach factories and foundry.

The wedding banquet was grand, as expected for the eldest daughter of a family of such wealth and high standing: twelve courses, starting with lobster bisque, followed by local river trout, toast and foie gras, a sorbet to precede the main course of pheasant with artichoke hearts accompanied by a variety of salads. The dessert course offered puddings, patisserie, cakes, ice creams and fruit, and each course was accompanied by a selection of wines, ending with a choice of liqueurs, sherry, port or champagne, and finally, coffee, cigars and cigarettes. The bridal couple sat side by side at a long table covered in white damask, decorated with wedding flowers, the family silver, china and glass; the servants of the house in their starched uniforms running to and fro, augmented by extra waiters from the town, each guest assigned their placement. The couple were flanked by their relations and facing the guests: Emma with her veil off her face now, her hair coiled up in the style of the times, Carl in his wedding fraque with a stiff white collar and fancy waistcoat, listening to a long list of toasts and speeches, everything as it should be, at least on the surface.

Except that the father of the bride was not present, a fact which did not escape the guests. Nor, of course, was the father of the groom, Pastor Jung – how to put it – having died some years previously. Then there was the fact that the mother of the groom was extremely large. And the puzzle that Frau Rauschenbach apparently was not present at the civil wedding, only the church wedding two days later. Above all there was the fact that the groom was a penniless doctor, not even a regular one but an *Irrenarzt*, a doctor of the insane, working in a Zürich lunatic asylum, marrying Emma Rauschenbach, one of the richest and most desirable young women in Switzerland.

Emma's wedding present to Carl was a gold fob watch and chain from the Rauschenbach factory, a magnificent piece of Swiss craftsmanship engraved around the upper inside rim with the words 'Fräulein Emma Rauschenbach' and 'Dr C. G. Jung' around the lower, with a charming design of lilies-of-the-valley at the centre, encircling two hands joined in matrimony, the date inscribed on a ribbon below: 16 February 1903. Not the 14th, then, but the 16th. Not the day of the civil wedding but of the church wedding. For Carl the church wedding was of little concern: he had rejected formal religion a long way back, during his battles with his father. 'The further away from church I was, the better I felt,' he wrote of his teenage self. 'All religion bored me to death.' But to Emma, raised in a conventional haut-bourgeois Swiss family, it was the church wedding that counted.

They spent the days before setting off on their honeymoon on 2 March at Ölberg. Early that morning they were borne away in the green everyday Rauschenbach carriage piled high with labelled trunks and hat boxes, Emma's mother and sister and the full complement of servants ranged on either side of the grand entrance, waving them off down the hill – '*Wünsch Glück! Wünsch Glück!*' – past the fountain and the vineyards and on to Schaffhausen railway station and the start of their journey – Emma in her travelling outfit and furs, Carl with his new set of tailored clothes and plenty of books.

After a short visit to Carl's mother and sister at the Bottminger Mill in Basel they travelled by Continental Express to Paris. The first-class carriages were arranged much like a drawing room, with heavily upholstered seats, writing tables and lamps for reading and curtains at the windows. The restaurant car was elegant, the menus all in French and the food excellent. The night was spent in lavishly adorned *salons lits*, with hand basin and mirror in the corner and the two bunks made up with fine bedlinen. Stewards in smart uniforms appeared at any time of day or night at the ring of a bell, and there was a quiet deference to all the proceedings. From Paris they travelled by train and boat to London, and thence down to Southampton to board the ocean liner taking them to Madeira, Las Palmas and Tenerife, always staying in the best hotels, then back via Barcelona, Genoa and Milan, arriving back at Schaffhausen on 16 April.

For Carl it was the start of a new life of considerable luxury; for Emma it was the beginning of getting to know the man she had married. He certainly looked the part: tall, handsome, beautifully attired. But at some stage during the honeymoon there was a quarrel about money. Swiss law gave the husband, as head of household, ownership of all his wife's possessions, and the power to make all final decisions in matters pertaining to family life, including the education of the children. Accordingly, the first thing Carl had done once they were married was to pay off his 3,000-franc debt to the Jung uncles. The amount of the debt tells you something about Carl's incipient taste for good living, a side of his character Emma was now discovering for herself. 'Honeymoons are tricky things,' Jung admitted to a friend years later. 'I was lucky. My wife was apprehensive – but all went well. We got into an argument about the rights and wrongs of distributing money between husbands and wives. Trust a Swiss bank to break into a honeymoon!' He laughed at the memory. But Emma had seen Carl's temper and how it could flare up out of nowhere. Ever since he was a boy Carl had flown into rages. In time Emma learnt that the rages soon passed. But for now it was distressing.

Then there was the question of sex. Was Emma surprised to discover that her husband was still a virgin? Two years earlier his mother had commented, on one of her visits to the Burghölzli and before he became engaged to Emma, that Carl knew too few women. Almost none, in fact. Jung himself later admitted he had not had 'an adventure before marriage, so to speak' and until his marriage he was 'timidly proper with women'. In one way this may have been a relief to Emma, given the terrible fate of her father. But she would not have been the only 'apprehensive' one, as Carl suggested. And now she knew Carl was not quite the man of the world he presented on the surface.

Then there was also the evident difference in their personalities. Quiet and studious as a child, as an adult Emma came across as shy and reticent, sometimes hiding behind the formality of her social background. With marriage she happily took her place in the background, giving those who did not know her the impression that she was, if not the *Hausfrau*, then just the wife. Carl was the extroverted one, the one everyone wanted to talk to. But it was only half the story, as Emma was finding out. The 'other' Carl, the hidden one, was unsure, introverted, plagued by a deep sense of social inferiority, and much in need of Emma's quiet support. Whilst Carl's childhood had been strange and lonely, Emma's had been safe, loving and happy, protected from the ills of the world at the Haus zum Rosengarten, overprotected perhaps. That world did not begin to break up until she was twelve, when her father first became ill with syphilis, and by then the confidence bestowed by a happy childhood was deeply embedded. She also had the confidence which came from belonging to the privileged *niveau* of society. It was a feeling Emma carried with her all her life, allowing her to behave in the quiet, modest manner noted by everyone who met her. A double confidence. Her new husband had neither.

On 24 April, shortly after returning from honeymoon, the Jungs moved into a rented apartment in Zürich, on

Zollikerstrasse, the road leading up to the Burghölzli asylum. Here, south of the city, the houses were solid rather than grand, the apartments spacious and the countryside within easy reach for afternoon walks. Unlike the eastern slopes of Zürich, which lay in the shade with damp air and bad soil, the southern slopes were on the sunny side, with good soil and little fog. Carl had taken up Bleuler's offer of a temporary post at the asylum, acting as his deputy whilst others were on leave, just filling in time whilst he looked for a better job elsewhere. It was a typically generous gesture from Bleuler who had already suffered Carl's precipitate resignation six months earlier, and it was received with a typically grudging gratitude by Carl.

With Carl back at work, Emma took stock of her married state. She started to set up home, arranging for furniture and fine antique pieces to be brought over by horse and cart from the Ölberg estate. Her mother came to help, plus a live-in maid, also from Schaffhausen, and gradually the apartment took shape. It was not the kind of living Emma was used to but it was only temporary after all. In many ways it was refreshing. From her bedroom window in the mornings she watched as bare-footed farm boys in short trousers and braces led cows down the hill to the slaughter-house, and smallholders from the neighbouring village of Forch came down in their wooden farm carts to sell their vegetables at the market, the husband in front driving the horse, the wife and children behind among the laden wicker baskets. Given Zollikon's proximity to the Burghölzli asylum, quite a few of the villagers worked there as helpers, the low wages still better than they could earn as farm labourers or domestic servants.

Carl left for work early every morning, walking up the hill through farmland and meadow, past the Burghölzli kitchen gardens where inmates in overalls were already hoeing and digging, watched over by the *Wärter*s in their long green garden aprons – then on through the gates of the asylum, up the front steps and straight into the early morning meeting. These days he was always smartly dressed in a suit and tie, waistcoat, and his gold fob watch and chain, which he wore till the day he died. But once on the wards he donned the doctor's white coat like the rest of the medical team and his daily routine began. He rarely got back before eight because Bleuler insisted Burghölzli staff eat their meals with the inmates, and the days were long for Emma left sitting in their apartment with nothing much to do and hardly any friends.

Some days she would make her way into the centre of town, walking down Zollikerstrasse, then taking the tram to Stadelhofen, where she joined a crowded, busy, modern city. Zürich was so much larger, noisier, and more commercial than Schaffhausen. Everything seemed so much faster. Trams could travel at fifteen kilometres an hour, conveying shop girls, hairdressers, waiters, bank clerks and office workers from the outskirts into the business district and back out again after a long ten-hour day. The tram drivers and conductors sported military-style uniforms and handlebar moustaches, as did the captains of the paddle steamers which plied their way across the lake, bearing freight as well as passengers.

By 1903 Zürich was a city on the make. Its population had doubled since 1890 to over 162,000 and was growing by the day. The unification of the outlying districts under one 'City Administration' in 1893 had propelled it into an era of thrusting capitalism and commercial activity, quickly followed by a property boom which led to massive land and building speculation, which in turn led to a financial crisis. But by the turn of the century the recovery was in full swing. Everything relied on Zürich's geographical position on the Limmat and the Sihl rivers flowing from Lake Zürich, their bridges linking the two parts of the town, with a chain of Alps beyond, each peak with a *Gasthof* or hotel on the summit. Zürich was postcard-pretty and good for business: commerce and finance, banks, a stock exchange, import and export, construction and engineering, heavy and light industry, all thrived on the shores of the lake. Further out in the industrial district of Sihl factories and sweatshops produced silk, cotton, lace and embroidery and bit machinery of every sort, all stored in warehouses on the bank of the Limmat beyond the Bahnhof Brücke, the bridge leading off Bahnhofstrasse. Rich foreigners were beginning to flock to Zürich for the Alps, the air, the shops and the lake. Fashionable hotels catered for this early tourist industry, none grander than the Baur-au-Lac, with gardens leading down to the lake. 'Hotel-keepers who wish to commend their houses to British and American travellers are reminded of the desirability of providing the bedrooms with large basins, foot baths, plenty of water, and an adequate supply of towels,' advised the Baedeker guide. Thomas Cook & Son had a bureau in the *Hauptbahnhof*, the central station, if further guidance were needed; they reminded travellers that no servants were allowed on the platforms: luggage had to be conveyed by the uniformed porters.

As spring gave way to a hot summer and she got used to her new surroundings, Emma found there was plenty to do in Zürich, quite apart from her usual reading, writing, needlework or knitting, and helping Carl to write up his daily reports and research. There were plenty of museums: the Ethnographical

Museum, the Municipal Museum of 'Stuffed Wild Animals', the Landesmuseum, a massive building in the neo-Gothic style recently opened in celebration of Zürich's increasing importance, housing an extensive collection of Swiss artefacts from prehistoric to modern times – bridal coffers, carved altars, old sleighs, tapestries, a distillery from the ancient Benedictine monastery of Muri, musical instruments, local Swiss *Trachts* (each canton having its own costume), china from the old factory at Schoren, banners and ducal hats and military uniforms of every kind, plus an entire Zürich house from the fourteenth century. But as far as Emma was concerned there was nothing more important than the Zürich Central Library with its collection of 160,000 books and its large, high-ceilinged, silent reading room – open in the morning from ten until twelve, and again after lunch from two till six, entry costing 60 rappen – ideal for her research into the Legend of the Holy Grail, which she now returned to with joy, like greeting an old friend.

Emma found a pleasing anonymity in Zürich: whilst the Rauschenbachs were the foremost family of Schaffhausen, known and respected by everyone, here Emma was mostly unknown and unregarded, in spite of her great wealth. The old ruling families of Zürich kept very much to themselves, entertaining one another in their villas in quiet, prosperous districts such as Seefeld, or meeting for private luncheons in fine restaurants, or at one of their many *Vereins*: the Yacht Club, the Rifle Club, the Rowing Club, the Riding Club, and the recently founded Automobile Club. Who among them would want to pass the time of day with the wife of an *Irrenarzt* employed at the Burghölzli asylum – an institution which had, after all, been built high up on the Zollikon hill, well away from the town, for a reason? This snobbery bothered neither Emma nor Carl, who were content, each for their own reasons, to be outsiders: Carl because he was born that way, Emma because society never held much importance for her, preferring as she did the company of close friends and family.

Marguerite often came to visit during that first year and the sisters would wander up and down the Bahnhofstrasse or the Parade Platz, though Emma bored of it quickly, one shop being much like the next. With the exception of the fashion houses. Both sisters were interested in the latest fashions, noting they were distinctly more modern in Zürich than sleepy Schaffhausen. Day wear was much the same: long-skirted and high-necked, worn with hats and button boots, but evening gowns were very décolleté, set off by heavy strings of pearls. The waists seemed narrower in Zürich, fishbone corsets pulled tight every morning by the maid, and the hats seemed larger, decorated with plumes and bows and veils, the hair curled with heated irons and pinned up with combs and grips. Feather boas and tassels were the fashion that year, and no lady went about without gloves, short for the day, long for the evenings. In the summer months the dark patterns gave way to white – long white skirts, high-necked white blouses, white stockings, white shoes. Swiss lace and embroidery came into their own then, along with parasols and wide straw hats decorated with ribbons and cherries and artificial flowers. Perfumes of violets, lilies-of-the-valley or eau de cologne were especially popular. On the Bahnhofstrasse in the recently opened *Salon de Beauté* there were boxes of rice powder, pots of white and rose creams, tiny bottles of nail polish and glass balls of bath salts displayed in discreetly half-netted windows, offering manicures and pedicures and other beauty treatments in the privacy of curtained-off niches, administered by trained young ladies in white overalls with their hair tightly pulled back.

After their shopping exertions Emma and Marguerite usually went to a café for *kaffee* or *thé citron* and *pâtisserie*. Other days they might take the cable tram up to the Grand Hotel Dolder, with its wide terrace and Zeiss telescope to better appreciate the view of the Alps. There were two picture houses showing newsreels and new features like *The Great Train Robbery*, jerky and silent except for the piano accompaniment, though good bourgeois people rarely visited them other than incognito, via a private side door. Emma and Marguerite preferred to promenade along the many *quais* which bordered the lake, or sit in the gardens of the Tonhalle – the grand concert hall built in 1895 in imitation of the Trocadero in Paris – listening to one of the military bands and talking about Marguerite's fiancé, Ernst Homberger, the son of another of their father's former business associates in Schaffhausen. One thing they did not do was visit the Panoptikum on the seamier side of the Bahnhof Bridge, with curtains drawn and dimly lit by gaslight, where men could, by looking through small panes, view a tiger hunt in the Sudan, or the 'Rape of the Sabines', or a wax Sarah Bernhardt in a negligee, or the 'medical' exhibition with lifesize models of naked women.

Sunday was Carl's day off. In the morning all the church bells, Protestant and Catholic, rang out across Zürich and families put on their Sunday best to attend the services, prayer books in hand, little girls with hair done up Heidi-style, boys in matelot suits. Not the Jungs, however. Carl had vowed never to set foot in a church again, other than on unavoidable occasions such as his own wedding, and he was determined not to have any of his children confirmed, remembering the debilitating boredom and depression which overcame him during the instruction given him by his father, who was all the while denying his own religious doubts, battling with his unnamed torments. Emma accepted Carl's rejection of formal religion without much trouble, not least because by Sunday they were often on the train to Schaffhausen, to be met at the railway station by coachman Braun as usual, then up the hill to Ölberg where Emma's mother and Marguerite and the servants waited, and Emma was back to the world she knew and loved.

Their first years of married life were happy for them both. A studio portrait taken at that time shows the couple side by side and close, looking into the lens of the camera with confidence and ease. Emma's wavy brown hair is done simply in a modern style – softly off the face. Her expression is shy but calm and direct, and quite determined. She is wearing a blouse in the Grecian style, high-necked, modern again. The skirt is long, simple and elegant, very unlike the elaborate outfits displayed in most of the fashion magazines. She may be standing on a box – the custom for studio photographs if the difference in height was too great – but she still looks half Carl's size. He stands there solid, with his hand nonchalantly in his pocket – dark suit, light waistcoat, a stiff-collared white shirt and small bow tie – a confident man of the world with no trace of the pompous and rather unhappy student of earlier photographs. His gold-rimmed spectacles glint in the studio lights and it is just possible to see the gold chain of his wedding present from Emma. His small moustache is trimmed the way it will be for the rest of his

life. In another take of the pose, Emma is smiling more, probably encouraged by the photographer, but Carl is just the same.
He looks – there is no other way to put it – like the cat that got
the cream.

'I'm sitting here in the Burghölzli and for a month I've been
playing the part of the Director, Senior Physician and First
Assistant,' Carl wrote on 22 August 1904 to Andreas Vischer, an
old friend from university days. It was summer, and everyone
else was away on holiday. 'So almost every day I'm writing
twenty letters, giving twenty interviews, running all over the
place and getting very annoyed. I have even lost another fourteen pounds in the last year as a result of this change of life,
which otherwise is not a bad thing of course. On the contrary,
all that would be fine (for what do we want from life more than
real work?) if the public uncertainty of existence were not so
great.' He might claim it annoyed him, but his tone is buoyant,
optimistic. There was nothing Carl loved more, he admitted,
than work – at least work that made sense to him: investigating
the dark corners of the mind. Emma was getting used to the idea
that she hardly ever saw him.

By the end of the year it was becoming clear that Jung was
not going to find another job elsewhere, certainly not in his
home town of Basel, which had been his plan. Apparently his
falling-out with Professor Wille had well and truly ruined his
chances: the job of director at the Basel asylum, which Carl had
applied for, was given to one of Wille's German colleagues, a
man named Wolff. Swinging from high spirits to low, Carl called
it the 'Basel calamity', which had 'wrecked for ever' his academic

career in Switzerland. Now he had to consider remaining in Zürich and at the Burghölzli. 'I might as well sit under a millstone as under Wolff,' he wrote to Vischer, 'who will stay up there immovably enthroned for thirty years until he is as old as Wille. For no one in Germany is stupid enough to take Wolff seriously, as Kraepelin has appropriately said, he is not even a psychiatrist. I have been robbed of any possibility of advancement in Basel now.'

Carl did not want to stay in Zürich nor take on a permanent post at the Burghölzli. He had no wish to do endless ward rounds, attend endless staff meetings, write up endless daily reports. He wanted to continue his research into the workings of the unconscious. He wanted to bring scientific proof to this new field and write scientific papers about his findings to be published in medical journals like the *Journal für Psychologie und Neurologie*. He wanted to make his name. But he had been rejected for the post in Basel and now he didn't know what to do. His crisis confronted Emma with the other Carl again, the one who fell out with colleagues through arrogance and an unshakeable conviction in his superior intelligence, but then became cast down and plagued with doubts when things went wrong. There was no steadiness in Carl, she discovered. And yet to the outside world he appeared his usual loud, confident, charismatic self. It was as though he were two different people.

The one thing which never left him was his vaulting ambition. 'As usual you have hit the nail on the head with your accusation that my ambition is the agent provocateur of my fits of despair,' Jung wrote to Freud some years later. He could admit it, but he could do little about it. Because the ambition was not just to gain worldly acclaim; it was in order to understand the complex workings of the unconscious because he suffered so much from it himself.

Bleuler again came to the rescue. He wasn't a crafty Swiss peasant for nothing and he wanted to keep his exceptional young colleague at the Burghölzli. So he made Jung an offer he couldn't refuse: he could continue the word association research he had begun with his Burghölzli colleague Franz Riklin, but more systematically, in a laboratory with proper equipment, and, under Bleuler's own supervision, write papers to be published in the medical journals. These could later be gathered into a book on this important new field of scientific research. Carl accepted. Bleuler and Jung shared an interest in the paranormal as a means of accessing the unconscious, and the Burghölzli already enjoyed an international reputation as great and controversial in its way as Freud had acquired after the publication of *The Interpretation of Dreams* in 1900. Everyone in the field, it seemed, wanted to research the unconscious, whether through dreams, hypnosis, the paranormal, or word association tests. Jung needed a base from which to launch himself. Bleuler needed the best doctors specialising in diseases of the mind, both for the sake of his patients and the reputation of the Burghölzli. Where else, after all, could Carl find so many guinea pigs to experiment on? Discussing it with Emma, Carl could see there was no better solution.

However, Carl was now tied to doing all the routine work as well – a twelve-hour day at least – and being completely teetotal, and, above all, having to move back and live on site. Nevertheless, by October 1904 Jung had exchanged the part-time for the full-time position, and he and Emma moved into an apartment in the main building of the Burghölzli, on the floor above Bleuler and his wife Hedwig and their children, bringing Emma's maid and the beautiful furniture with them. The only concession made to Herr and Frau Doktor Jung was that Carl was allowed to miss the midday meal with patients and staff and return to the apartment to have lunch with his wife. So began Emma's new life as the wife of an *Irrenarzt*, living amongst hysterics, schizophrenics, catatonics, alcoholics, addicts, chronic neurotics and suicidal depressives – people who had lost their minds for one reason or another, and who spat and screamed and paced the wards, up and down, shouting obscenities, tearing at their hair, breaking the furniture. The contrast with her former life was complete.

Now she had to find a way of relating to the kind of people she had never expected to meet: inmates and staff alike.

It is telling that Emma was able to do it quickly and naturally, as testified by everyone who met her. Of course they knew Herr Doktor Jung had married a wealthy wife, but it was soon noticed that the Frau Doktor did not 'act wealthy', dressed simply, and was friendly to everyone equally, regardless of position. She was quite reserved, it was true, but she was liked. For Emma the problem was more the question of how to fill her days. The maid did the cooking and the cleaning, and, unlike Hedwig Bleuler as the wife of the Herr Direktor, she did not have a formal role in the institution. She still helped Carl privately with his research and reports, and she continued to go to the library in Zürich and research the Legend of the Holy Grail, but it still left her with many hours alone. There were just two people nearby who she could visit regularly – a duty or a pleasure, depending on how you looked at it: Carl's mother and his sister Trudi, because once it became clear that Jung was staying in Zürich he swiftly moved them out of the Bottminger Mill and into an apartment in nearby Zollikon. He did not have much time to see them himself so the visits naturally fell to Emma, the wife and daughter-in-law. It can't have been easy: most of the time her mother-in-law was normal and good company, but she still heard voices, had visions, and made sudden prophetic announcements. Her second personality always hovered in the background, and you never knew when it might emerge.

Soon Emma found a role to play at the Burghölzli after all, because Eugen Bleuler believed women and wives should be included as much as possible in the daily life of the institution – another of his progressive ideas, quite alien to other lunatic asylums where all women other than the female *Wärters* were kept well away. Emma was encouraged to join in, participate in social events, and sit at the same tables as the patients as long as there was no danger. There were also regular evening discussion 'circles' to which all doctors and their wives were invited. And if a patient was in the rehabilitation phase of their treatment and they were allowed out of the asylum grounds, it was as likely to be Frau Jung or Frau Direktor Bleuler as much as it was one of the *Wärters* who accompanied them on walks, or even on a short shopping trip. 'The years at Burghölzli were the years of my apprenticeship,' Jung acknowledged later. But in a smaller way they were Emma's too. 'Dominating my interests and research was the burning question, "What actually takes place inside the mentally ill?"' Jung wrote. Now, almost by osmosis, it came to interest Emma too.

The other plus for Emma at the Burghölzli was Hedwig, who was a remarkable woman and an early feminist. Eugen Bleuler was already forty-four when he married her in 1901; Hedwig was then a history teacher, twenty-four years old, clever, elegant, charming. They met at one of Bleuler's lectures at Zürich University, and wed within the year. Hedwig's ambition had been to become a lecturer herself but it was not possible in Switzerland at that time, it being the preserve of men. And once she was pregnant she could not continue her studies anyway, so her life centred round her role as Frau Direktor Bleuler, and she made very good use of it, supporting her husband in every way she could. Their apartment was large enough to accommodate two maids, had bedrooms enough for their growing family, and a large, sunny room for Bleuler's schizophrenic sister Pauline. There was a library and a stairway which led directly down to the wards, so Bleuler could be there at any time of day or night if there was a crisis. Hedwig's special interests were the abstinence movement and women's position in a modern society. Years later she travelled throughout Switzerland lecturing on both, but whilst the children were growing up she limited herself to helping her husband at the Burghölzli and editing his lectures and published papers. It was a happy marriage but she always regretted having to give up her own career. 'The woman is always the one who has to make the sacrifices,' she said. Another clever woman, then, who had to find her intellectual satisfaction through her husband's work.

Once back full time at the Burghölzli, Jung was soon dominating the place with his energy and booming laugh. He was happy to spend hours with a patient listening to their strange utterances, finding meaning in their madness. While he was liked by many, some of his colleagues found his presence irritating, accusing him of throwing his weight around, doing just what he wanted and not bothering with the dull administrative work. The *Wärters*, who did all the day-to-day caring and were only allowed home one afternoon a week, claimed the doctors spent more time on their research than on routine ward rounds, especially Herr Doktor Jung. But Bleuler let him get away with it. Auguste Forel, the former director who still liked to wander about the institution, was soon asking: 'Who's running this place, Bleuler or Jung?'

The reason was simple: Bleuler did not want to lose Jung. So he allowed Jung and his 'esteemed colleague Herr Doktor Riklin' to spend hours pursuing their word association tests in a laboratory at the back of the main building, near both the laundry and the dairy, wafts of steam emanating from the one and the moos of cows from the other. Riklin was the junior partner in the venture but he had worked on word association tests before, in Germany with Gustav Aschaffenburg who followed Dr Wundt, who in turn referred back to the first tests conducted by Galton in England in the 1880s, each contributing to the growing body of evidence about the unconscious. All over Europe doctors in asylums were experimenting on their patients, and Jung's papers on the word association tests are filled with references to others already embarked on the same track. The difference was that Jung and Riklin's methodology was far more systematic, more scientific. They published their findings in the *Journal für Psychologie und Neurologie*, and Bleuler wrote the introduction when the book was published in 1905 as *Studies in Word Association*.

The method was described as 'the uttering of the first word which "occurs" to the subject after hearing the stimulus-word',

and suggested that it would help in the diagnoses and classification of *dementia praecox*, epilepsy, various forms of imbecility,
some forms of paranoia, and the diseases grouped under hysteria, neurasthenia and psychasthenia, 'not to speak of manic
depression with its well-known flight of ideas'. Jung and Riklin
went about it systematically and scientifically, by using a much
larger pool of guinea pigs and by increasing the number of stimulus words from 100 to 400 – nouns, adjectives, verbs, adverbs
and numerals, using the Swiss dialect form when necessary –
and timing the reactions with the greatest possible accuracy
using the 'one-fifth-second stop watch', the very latest model
from the Rauschenbach factory in Schaffhausen. The reaction
times varied considerably, sometimes taking up to 6 seconds.
Galton had found that the average time was 1.3 seconds. Jung
with his modern stopwatch found it was 1.8 seconds, women
subjects taking slightly longer. To distract the conscious mind a
metronome was used. Other researchers used whistles, trumpets
or darkened rooms. At times the tests were done when the
subject was 'in a state of obvious fatigue', and once 'in a state of
morning drowsiness' (Jung himself). One important innovative
decision of Jung's was to test 'normal' as well as 'abnormal'
people, and the groups were divided into male and female,
'educated' and 'uneducated'. Everything had to be written up,
by hand, night after night. Emma was kept busy, and she was
happy to do it.

In those early years of psychoanalysis it was common and
considered perfectly acceptable to use friends and relatives as
well as patients as subjects for scientific research. Jung and
Riklin's first test was on 'Normal Subjects': thirty-eight persons
in all, nine men and fourteen women of whom were classified as

'educated'. The rest were mostly male and female *Wärters* at the Burghölzli, who, Jung declared, were not so much 'uneducated' as 'half-educated'. The tests were often repeated so that in the end they had 12,400 associations and timings on which to base their conclusions. There were statistics and tables and graphs, and a complicated list of different types of reaction, including those which showed evidence of 'repression', a term made familiar in the writings of Sigmund Freud. As with Freud, there was often enough of a description of the person to make an enticing story. Thus 'Uneducated Woman, Subject No. 1: she is of country origin and became an asylum nurse at the age of seventeen, after having brooded at home for over a year over the unhappy ending of a love affair'. This subject would not or could not understand the stimulus words 'hate, love, remorse, rattle glass, hammer ears . . . because they intimately touched the complex which she was trying to repress'. The term 'complex' was coined by Jung and Riklin to denote 'personal matter . . . with an emotional tone'. It could be spotted by a significantly longer reaction time and the peculiarly forced nature of the reaction.

Then there were *clang* associations – that is, based upon sound rather than on concepts: simple, thoughtless, sound similarities – and interesting 'preservation' ones, first noted by Aschaffenburg, where 'the current association revealed nothing, but the succeeding one bears an abnormal character'. There was also the egocentric reaction: grandmother/me; dancing/I don't like; wrong/I was not, and so on. 'If we ask patients directly as to the cause of their illness we always receive incorrect, or at least imperfect information,' explains Jung. 'If we did receive correct information as in other (physical) illnesses, we should have known long ago about the psychogenic nature of hysteria. But it is just the point of hysteria that it represses the real cause, the psychic trauma, forgets it and replaces it by superficial "cover causes". That is why hysterics ceaselessly tell us that their illness arose from a cold, from over-work, from real organic disorders.' He compared their method with Freud's 'free association' and reminded the reader that a 'delicate psychological intuition in the doctor is as much a requisite as [is] technique for a Psycho-Analysis'. If this all sounds familiar now, it was not then, back in 1904.

Emma appears as 'Subject no. I, aged twenty-two, very intelligent', in the 'educated' category. By way of introduction Jung wrote: 'No. I is a married woman who placed herself in the readiest way at my disposal for the experiment and gave me every possible information. I report the experiment in as detailed a way as possible so that the reader may receive as complete a picture as possible. The probable mean of the experiment amounts to 1 second.'

The first five word associations were: head/cloth, 1 second; green/grass, 0.8 seconds; water/fall, 1 sec; prick/cut, 0.8 seconds; angel/heart, 0.8 seconds. So far so good, but reaction 5 was deemed striking because 'the subject cannot explain to herself how she comes to *heart* . . .'. Emma denied it was the result of any disturbance from without, and could not find any inner one either. Jung concluded it might therefore be some unconscious stimulus, very likely one of Aschaffenburg's 'preservation' reactions, carried over from the previous prick/cut, which caused 'a certain slight shade of anxiety, and image of blood'.

'The subject is pregnant,' Jung noted with scientific detachment, 'and has now and then feelings of anxious expectancy.'

There followed a sequence of stimulus words which were not especially memorable, except that 'to cook' endearingly elicited 'to learn', as did 'to swim' – because it was her sister Marguerite who was the swimmer, whereas Emma was apparently still learning. Only Carl could know why 28: lamp/green took 1.4 seconds. It followed threaten/fist, and he noted it was clearly another case of 'preservation' and that lamp/green denoted her home life (the colour of the lampshades). He does not say which home, Zürich or Schaffhausen, but it was most likely Schaffhausen and the fist her father's as he became more and more ill and desperate.

Further associations offered little which was significant for the test but tell us something of Emma's outlook on life: evil/good; pity/have; people/faithful; law/follow; rich/poor; quiet/peaceful; moderate/drink; confidence/me; lover/faithful; change/false; duty/faithful; serpent/false. And then we come, in no particular order, to: family/father and mother/tell and dear/husband. Father is still the head of the family. Her mother is the one to whom she tells almost everything. Her husband, she loves.

Using his intuition, and inevitably his personal knowledge of Subject No. I, Jung focused on a sequence of associations, nos 70–73: blossom/red; hit/prick; box/bed; bright/brighter. The first pair only took 0.6 seconds. 'She explains this short reaction by saying that the first syllable of the stimulus-word Blo-ssom brought up the presentation of blood. Here we have a kind of assimilation of the stimulus-word to the highly accentuated pregnancy complex . . . It will be remembered that in the association Prick/Cut (no 4) the pregnancy complex was first encountered,' yet:

Box/Bed which followed Hit/Prick went quite smoothly without any tinge of emotion. But the reaction is curious. This subject has now and then paid a visit to our asylum and was alluding to the deep beds used there, the so-called 'box-beds'. But the explanation rather surprised her, for the term 'box-bed' was not very familiar to her. This rather peculiar association was followed by a clang-association (Bright/Brighter) with a relatively long time . . . The supposition that the clang-reaction is connected with the previous curious reaction does not, therefore, seem quite baseless . . . assuming a clang-alteration at the suppressed pregnancy complex, the complex becomes very sensible.

By Association 83 we come to injure/avoid. The German for 'injure' is *schaden*. As Jung explains, the word is very like and easily confused with *scheiden* – divorce. Luckily, apart from these, there were plenty of happy associations too.

So the main complex which emerged from the test was Emma's fear of the pain of childbirth. Association 43, despise/*mépriser*, reminds us how well Emma spoke French, but this, it turns out, was not the point. The reaction took 1.8 seconds. Why so, Jung wondered? 'Despise is accompanied by an unpleasant emotional tone. Immediately after the reaction it came to her that she had had a passing fear that her pregnancy might in different ways decrease her attractiveness in the eyes of her husband. She immediately afterwards thought of a married couple who were at first happy and then separated – the married couple in Zola's novel *Vérité* [Truth]. Hence the French form of the reaction.' Poor Emma.

In his summing up of 'Subject No. I', Jung commented:

In reality our experiment shows beautifully, the conscious self is merely the marionette dancing on the stage to a hidden automatic impulse . . . In our subject we find a series of intimate secrets given away by the associations . . . We find her strongest actual complex to be bound up with thoughts about her pregnancy, her rather anxious expectancy, and love of her husband with jealous fears. This is a complex of an erotic kind which has just become acute; that is why it is so much to the front. No less than 18% of the associations can certainly be referred to it. In addition there are a few complexes of considerably less intensity: loss of her former position, a few deficiencies which she regards as unpleasant (singing, swimming, cooking) and finally an erotic complex which occurred many years back in her youth and which only shows itself in one association (out of regard for the subject of the experiment I must, unfortunately, omit a report on this).

Jung ended the German text of *Studies in Word-Association* giving 'special thanks to Frau Emma Jung for her active assistance with the repeated revision of the voluminous material'.

5

Tricky Times

Agathe Regina was born at Emma's family home in Schaffhausen on 26 December 1904, a Christmas child. Emma was twenty-two, had been married for a year and ten months, and now she was a mother. Overnight, it seemed, her life had changed again, this time for ever.

It was common for mothers of Emma's social class to use a wet nurse to still their babies. But Emma breastfed hers herself. She did not want to hand her baby over to a stranger and deprive herself and the child of this pleasure, as Carl later described it to Freud. Nevertheless, a first baby is a strange and unknown experience. Emma had plenty of help with 'Agathli' – little Agathe – whilst she was still at Ölberg with her mother and sister Marguerite, but when she moved back to Zürich and the Burghölzli, keen not to be away from her husband for too long, she largely looked after the baby herself. Feeding took many hours. Waking through the night disturbed sleep. Changing the thick towelling nappies was onerous, though it was the maid's job to soak them in a pail before washing them by hand, then through the wringer, then hanging them out on a wooden horse by the tiled stove to dry. It was a winter with deep snow, white pitched roofs with icicles hanging from the eves and the window sills, and no hope of using the washing line in the garden. There were coughs and colds and nappy rashes to deal with, and the terrible responsibility of a new life. For a young mother who had shown such anxiety about the pain of giving birth and the fear that having a baby might affect the feelings her husband had for her, Emma was having to grow up very fast indeed.

In addition, Emma had just started what she had longed to do since leaving school: furthering her education. Though it was done circuitously, by helping her husband with his work, it was certainly an education. And she was still going down to Zürich Central Library to pursue her own research into the Grail legend. But now Carl and Emma slipped back into more conventional roles: she as wife and mother; he as husband, breadwinner and authoritative head of house. Carl did not subscribe to '*Das Weib sei dem Mann untertan*', 'The woman shall be subservient to the man', as a popular book, *The Way to the Altar*, quoting the Bible, reminded, but once Agathe was born things changed in the Jung household. Carl was conventional when it came to parenthood: he saw children largely as the responsibility of the mother. Now his sister Trudi came to help with the secretarial work and Emma found herself more and more on the sidelines. If she hoped it was merely temporary until she found her feet, it didn't work out that way: within six months she was pregnant again. Anna Margaretha, known as Gretli, and later Gret, was born on 8 February 1906.

Emma's story was typical enough of the times. Many women were frightened of getting pregnant, either because they already had too large a family or because they were not married, which was worse. There was no safe method of birth control, and a reliance on coitus interruptus had limited success. In 1909

Richard Richter, a German doctor, would develop an early form of an interuterine device using the gut of the silkworm, but it was not marketed until the 1920s, by which time Marie Stopes had opened the first birth control clinic in England, and there were similar initiatives in America and France and Germany. The Catholic Church, however, forbade any form of contraception and the Protestant Church did not look on it with favour either, believing that it was the duty of Christian marriage to 'increase and multiply and fill the earth'. It all came too late for Emma and Carl anyway.

Sigmund Freud could have told them. His letters to his doctor friend Wilhelm Fliess are full of pleas that he come up with a reliable form of contraception. Every month he and Martha worried she might be pregnant again. Fliess looked into the 'rhythm method', making calculations and trying to work out a safe period during Martha's monthly cycle; he even tried to come up with a similar cycle for Freud himself. Evidently it was not successful: Martha became pregnant six times in ten years. For the first three months of her sixth pregnancy she insisted it was the start of the menopause, not another pregnancy. She couldn't face it. She had never wanted more than four children. Once the baby, Anna, was born, Martha went to her mother's in Wandsbek for several weeks of recuperation and there developed 'a writing paralysis': she found she literally could not form the words to write Sigmund a letter. Her face was puffy and her teeth hurt. She was only thirty-four and she was exhausted. After that they stopped having sex, that side of their marriage becoming 'amortised', as Freud later confided to Emma.

* * *

In March 1905 Emma Jung's father Jean Rauschenbach died. Marguerite and Ernst Homberger, who had married a year earlier, now moved in with Bertha whilst they built a house of their own. Ernst took over the running of the Rauschenbach business from Bertha and Jean's sister, though the two women had managed it all splendidly during the last years of Jean Rauschenbach's decline. But custom had it that if there was a man in the family there was no need for the women to carry on working. Ernst, a quiet, ambitious, tough man twenty years his wife's senior, ran the business for the rest of his working life.

The immediate reaction to Herr Rauschenbach's death was relief as well as sorrow, his last years being so terrible, shut away from the world. 'A poor rich man finally closed his eyes last night, but their light had already gone out a long time ago,' stated the obituary in the *Schaffhausen Tagesblatt* on 3 March 1905, written by an old friend. 'Herr Johannes Rauschenbach, manufacturer, died at his country seat at Ölberg, aged 48. He saw nothing of his beautiful new home, being already fully blind by the time work on it was begun . . . He was no longer able to see his two young daughters grow into lovely young women, nor was he able to lead them up the altar into married life, because by then he had already been leading the life of a hermit for many years, shut away in his upstairs room.' He was described as the son of a simple locksmith who became the founder and owner of the world-famous machine factory and foundry, after the early death of his father. After fetching himself 'a beautiful young bride from Uhwiesen . . . the world stood gloriously before him. He threw himself into the business which his father had built up, having been underestimated like the sons of all self-made men. And he felt it, resulting in a certain disdainful sarcasm.' But he did well, overcoming a financial crisis in the Budapest branch of the family business. Then 'the most terrible bad luck overcame him'. At first, when he started to go blind, 'he was almost totally cast down. In vain was the loving care of his wife! Heart-breaking was his despair.' Over the years he became calmer, and finally he was at peace from his suffering. 'His life was like a glorious morning, a sunny midday, but a heavy, rainy evening. Let us remember him as he was in the morning and midday. Rest in peace.'

Herr Rauschenbach left all his worldly goods and his vast fortune to his wife and his two daughters – in effect, to his two sons-in-law. From now on Ernst remitted an annual sum from the profits of the business to Carl, who duly sent an annual letter of thanks. It made Emma and Carl even wealthier, so wealthy in fact that Carl would never need to work again. He could leave the Burghölzli any time he wished.

But he didn't wish. His 'real creative work', as he put it, was at the Burghölzli, doing research on word association in the laboratory. There was an outpatients' clinic now too, so his subjects, 'normal' and 'abnormal', could be drawn from these as well as from the inmates and the staff of the asylum. Where else could Carl find such a pool of 'volunteers'? Bleuler allowed him private patients as well, another concession which annoyed some of his colleagues. When in 1905 Karl Abraham arrived from Germany to join the staff, together with assistants Hans Maier and Emma Furst, Jung was appointed senior physician. He also became a guest lecturer in psychiatry at Zürich University, following in Bleuler's footsteps.

From the start Jung knew how to attract an audience to his lectures, avoiding the dry academic style of most professors, including Bleuler, and instead peppering his presentations with dramatic stories, instructive case details and sharp insights. 'My aim was to show that delusions and hallucinations were not just specific symptoms of a mental disease but also had a human meaning,' he said. 'To my mind, therapy only really begins after the investigation of that wholly personal story. It is the patient's secret, the rock against which he is shattered. If I know his secret story, I have a key to the treatment. The doctor's task is to find out how to gain that knowledge. In most cases exploration of the conscious mind is insufficient. Sometimes an association test can open the way; so can the interpretation of dreams, or long and patient human contact with the individual.' Secrets and rocks. Rocks and secrets. Jung knew it all from personal experience. But for those who came to his lectures it was all thrillingly new.

There was no shortage of stories: the seventy-year-old female inmate of the Burghölzli who had been there for fifty years, bedridden for the past forty. She couldn't speak. She ate with her fingers. She made curious rhythmic motions with her hands. At Burghölzli demonstrations she was presented as a hopeless case of catatonic *dementia praecox*. But one day a female *Wärter*, who had worked at the Burghölzli for decades, told Jung that in the early years the woman had been put to work by Bleuler making shoes, which kept her busy and her mind occupied. Not long after Jung arrived at the asylum the woman died. A brother came to the funeral. He told Jung his sister had once been in love with a shoemaker but it had not worked out. Now Jung understood the rhythmic motions with her hands. 'That case gave me my first inkling of the psychic origins of *dementia praecox*,' he wrote.

Another case: a female doctor, who was one of Jung's private patients, had murdered her best friend out of jealousy because she wanted to marry the husband. She was never found out and she married the widower, but he died young. She had little contact with their only daughter. People and even animals seemed to turn away from her. She lived 'plunged into unbearable loneliness'. Finally she came to see Jung, to 'confess'. He did not give her secret away and she left to carry on her life, but he dreaded what fate held in store for her.

'He kept his students spellbound by his temperament and the wealth of his ideas,' recalled Ludwig Binswanger, who came as a volunteer to the Burghölzli in 1906 and to write his dissertation under Jung's supervision. But it was not only Jung's students who were spellbound: the lectures were open to the public and soon he had a large following of wealthy Zürich women who had too much leisure on their hands. Here they found an outlet for their underused intellects and often for their underused emotions too, what with talk of erotic complexes, repression, and this new world of the unconscious, illustrated with fascinating references to literature, mythology, art and plenty of real-life stories. Carl's lectures were soon the best act in town and the medical students complained that they often could not find a seat in the lecture theatre, so popular were they with the so-called *Pelzmäntel* – fur-coat – brigade.

For Carl it was a revelation. He had never been a 'ladies' man'; he had, in fact, held himself somewhat aloof from that kind of thing, mostly in self-defence, lacking the confidence and the know-how to woo a young woman. He never thought of himself as handsome, and in truth the 'Steam-Roller' was not very handsome, nor confident, nor charming, being too loud and brash, peasant-crude at times, and suffering from a keen sense of social inferiority. But now, three years into marriage with Emma and approaching thirty, everything was changing. His new wealth made him confident, he had lost some weight and he looked handsome in his expensive clothes worn casually *à l'anglaise* like the gentlemen he had met in Oxford. His position at the Burghölzli was a good one, his research was consuming, his lectures a sell-out, and he was doing court work too – giving expert psychiatric evidence in legal cases, which allowed him to take time off and go down to Zürich, lunch in a restaurant or go for a walk along one of the *quais* before taking the tram back up the hill again. Underpinning it all was his domestic life – a happy one, from his side anyway – with Emma and their two little girls.

For Emma things were not so good. Where Carl's life was full of variety, hers was limited. Where he was getting on with his career, she, the 'very intelligent' subject of the word association tests, felt stuck. Where he was free to come and go, she felt trapped. It was not that Emma was unhappy to be a mother, just held back. And in fact there were always happy times too. They talked together. They laughed together. They enjoyed their children together. A series of charming, fuzzy snapshots taken with a box camera in the Jungs' apartment at the Burghölzli shows Carl seated on the canapé with Agathli and Gretli, happy in the role of Papa, clothes awry. Another is of Emma, on the same canapé, holding Gretli, smiling and demure. Jung took the one of her, she took the one of him. They are informal and domestic in a way rarely seen in photographs of the time, and they show a very different side of Carl. Emma loved her children. It was just that, as Hedwig

Bleuler said, it was always the woman who had to make the sacrifices. Never the man.

Underlying everything was Emma's anxiety that her husband was beginning to attract so much female attention. Rumours of the admiring *Pelzmäntel* ladies at the university lectures did not take long to reach the Burghölzli. And Carl himself told Emma about some of his female patients who were caught in an 'erotic complex', falling in love with their doctor, a common hazard of the profession.

He had been treating two female inmates, both aged twenty-four, both diagnosed as 'hysterics'. The first had been institutionalised because she was overcome by a debilitating exhaustion, lying in the house all day or down in the cellar if it got too hot in the summer. Her symptoms had started when she was still a schoolgirl, with attacks of tremors in her right arm which spread to the whole body, accompanied by tics and shrieks. By the time she came to the Burghölzli she thought she was going mad. Applying his word association test and analysing her dreams using Sigmund Freud's methods, Jung uncovered a sexual complex and suspected sexual abuse, probably incest. Gradually her symptoms improved. She was enjoying the one-to-one attention she got from Herr Doktor Jung and as time went on she became more and more demanding. Finally, Carl found he had to 'pitilessly disturb her illusions' about himself. 'It is to no purpose that she seeks to still her craving for love by falling in love with her doctor,' he wrote, 'for he is already married.'

Her friend, 'Miss L' in Room 7, had become ill through an unhappy love affair and was another 'who gushes about the writer. Like the patient, she has become ill through an erotic complex.' Transference was not well understood in the early years of psychoanalysis, nor the position of women in a society where marriage was virtually their only outlet, and doctors like Jung and Freud often encountered such fixations. Carl had first come across it with his cousin Helly: 'The girl had of course fallen hopelessly in love with me,' as Carl later wrote. 'But I paid very little attention to it and none at all to the role I was playing in her psychology.'

Jung, previously so awkward with women, was beginning to enjoy the attention. Gossip and rumour, so much part of institution life, abounded, and Emma could not fail to hear it, though she probably knew most of the talk was tittle-tattle not to be taken seriously. However, on the night of 17 August 1904 a young woman had been admitted to the Burghölzli as a case of emergency, and this one would be serious.

It was 10.30 p.m. and Jung was the doctor on duty, so he was the one to write the admission notes. A medical police officer and an uncle had brought the young woman in, having taken her by force from the Hotel Baur-en-Ville, where she had been staying with her mother. She had already been thrown out of the Heller Sanatorium at Interlaken because of her behaviour. Jung noted that she was laughing one moment, crying the next, 'in a strangely mixed compulsive manner', and that she had numerous tics, rotated her head in a jerky fashion, stuck out her tongue, and that her legs were twitching. She kept insisting she was not mad, she just had a terrible headache and got 'upset' at the hotel. Jung recorded the uncle's statements as 'meagre and evasive', partly because his command of the German language was not good, being 'an old Russian Jew'. Apparently Fräulein Spielrein had always been 'rather hysterical', and had been 'ill' for the past three years. When she was thrown out of the Heller Sanatorium she was meant to go to the Monakow clinic but they refused to take her 'because she was too disturbed'.

Jung's report was detailed, to the high standards required by Bleuler, and he did not get back to the apartment until close on midnight.

The next day Jung carried out a full anamnesis. The young woman was Sabina Spielrein, the nineteen-year-old daughter of wealthy Russian Jews from Rostov-on-Don. The female *Wärter* who had spent the night in her room reported that the new patient had been anxious in the dark and wanted a light, that she said she had two heads and her body felt foreign, but otherwise it had been a quiet night. The examination was difficult – 'like walking on eggshells' – and it became 'a powerful battle' between doctor and patient. When Jung pointed out that she would have to divulge 'everything' if she wanted to get better, Fräulein Spielrein replied 'she would and could never talk about it, and in any case does not want to be cured at all'. Jung noted 'stomach pains, angina a thousand times, precocious, sensitive'.

Sabina Spielrein was the eldest of five children. Her only sister, whom she had loved 'more than everything in the world', had died aged six of a stomach complaint. Her father was a successful businessman. At the age of five Sabina had unac-countably been sent to a school over a thousand kilometres away in Warsaw; Jung's report did not say if her mother went with her. By the age of six she already spoke German and French. Later she was taught at home until she entered the *Gymnasium*, where she was considered 'intellectually very advanced'. She complained that the teachers were 'very stupid'. She played the piano and sang. She was interested in the natural sciences and wanted to study medicine at Zürich University, which already had some wealthy Russian female students, part of a Russian colony in Zürich which included Jews escaping the pogroms.

Jung arranged for Fräulein Spielrein to be accommodated in the women's section with a female *Wärter* and a private room to herself, costing 1,250 Swiss francs a quarter, a sum roughly equivalent to Jung's salary for a year. He wrote to Fräulein Spielrein's mother at the Hotel Baur-en-Ville, requesting Sabina's belongings to be sent over, and signing it 'Doktor Jung, Senior Physician'. For now, the mother remained at the hotel and the father returned to Rostov.

Over the next few days Emma noticed that her husband seemed even busier and absorbed than usual. He spent several long sessions with Fräulein Spielrein, battling to get what he called a 'confession' out of her. Spielrein admitted that her father hit her several times on her bare buttocks, sometimes in front of her siblings, and when he hit her she had to kiss his hand afterwards. When she talked about this, 'innumerable tics and gestures of abhorrence occurred ... which have a certain connexion with her complexes'. Jung noted that her father was always threatening suicide while at the same time tyrannising the family. One brother had hysterical fits of weeping, another suffered from tics; the last was described as a 'melancholic'. Fräulein Spielrein said she loved her father 'painfully', but she could not turn to him. 'The peak of the experience,' wrote Carl, 'was that her father was a *man*.' Her mother also beat her, in front of her brothers and even their friends.

Sabina Spielrein's reaction to all this was to chastise herself, and fall in love with her uncle, and later with an assistant doctor at the Heller Sanatorium. When she was fifteen she tried to starve herself to death 'because she had made her mother angry'. Aged seven, she had become very pious and started to talk 'with a spirit'. God spoke to her in German. She felt she was 'an extraordinary person'. All this was a bit too close to the bone for the 'other' Carl, the one who had heard voices, had visions, and knew he was 'extraordinary'.

Whilst her husband worked all hours, Emma's life gradually settled into a pattern. During the week she spent time with Hedwig Bleuler and her children, and helped her with social events for the asylum inmates. In the evenings, once the baby slept, she worked on Carl's research and daily reports. Weekends were often spent in Schaffhausen with her mother and Marguerite and Ernst, where Agathli received the happy attention Emma herself had enjoyed as a child. But she did not like to be away from Carl for too long. There were growing rumours about Fräulein Spielrein, and Emma could not fail to notice that this private patient was more demanding than all the rest. Still, that spring, Carl and Emma were able to go for a short holiday to Berlin, leaving the baby in the care of the children's maid, again helped by Groma Jung and Trudi. For once Emma had Carl to herself, in the great city: the palaces, art collections, the avenue of Unter den Linden, the Brandenburg Tor, the cafés and the restaurants, the theatres, the opera, though Carl spent much of the time in bookshops and visiting hospital laboratories. But, perhaps at Emma's insistence, it began a ritual which lasted for the rest of their lives: an annual spring holiday on their own, just the two of them.

It would have been easy to feel cut off from the rest of the world, living behind the high walls of an institution like the Burghölzli, but Carl and Emma were both minded to keep abreast of events. They took three newspapers: the Swiss *Neue Zürcher Zeitung*, the French *Le Monde* and the English *Daily Telegraph*, to be well informed internationally as well as locally, and to practise their languages. They also took the *Schweizer Illustrierte* magazine and *Nebelspalter*, a Swiss satirical publication; and *Punch*, the *Illustrated London News* and *L'Illustration* – all passed on from Bertha via the Lesezirkel Hottingen, the fashionable reading circle in Zürich.

The *Neue Zürcher Zeitung* felt there was too great an appetite for change and everything was going too fast. Every week there was a report about some tram or train accident, and did the telephone and telegraph systems really improve the average Zürich burgher's life? The rich were merely getting richer. 'An automobilist is a capitalist gone mad,' agreed the editor of *Der Bund*. 'Who sits at the wheel of this appalling vehicle? A speed fanatic, an addict of self-indulgence, and a show-off!' Internationally the talk was all about revolution and Russia: the peasants were starving, workers were striking, anarchists and socialists agitating against a dictatorial Tsar. Now the strikes were spreading across Europe, even reaching Zürich where the workers, usually so docile, demanded a nine-hour day and better working conditions. At the Arbenz automobile factory there was a strike after one of the workers was dismissed for 'sabotage'. The employers got rid of the 'trouble-makers' and advertised for replacements, adding: 'Only hard-working, quiet and solid workers will find full employment,' When the unrest in the streets got out of hand the military was called in to help the city police and many arrests followed. The good burghers of Zürich had never seen such a thing. Worse was to follow: the San Francisco earthquake of spring 1906 would almost ruin the Swiss insurance business, and the New York stock exchange crisis would do the same for the banks.

At the Burghölzli, Carl acquired another private patient: an American, Elizabeth Shepley Sergeant. She arrived at the asylum after suffering a breakdown whilst travelling in Europe with an aunt, becoming the first in a long line of wealthy Americans to cross the ocean to be treated by Herr Doktor Jung. The Burghölzli up on Zollikon hill might seem cut off from the rest of the world behind its high walls, but it was getting better and better known in the wider world. At the same time as Jung was attracting attention, a book by the Burghölzli's previous director, Auguste Forel, was causing quite a stir. *The Sexual Question* was 600 pages of shocking 'progressive' ideas, championing the equality of the sexes and advocating no punishment for 'mutually consensual homosexuality', as well as promoting teetotal abstinence. The Churches – Protestant and Catholic alike – were up in arms and the book was banned in parts of Switzerland. Internationally, however, it was a runaway success.

Meanwhile Fräulein Spielrein had been formally diagnosed with 'hysteria', and her treatment was taking up even more of Jung's time, with sessions lasting up to three hours. He noted that Fräulein Spielrein 'felt as if someone were pressing in upon her, as if [someone], something was creeping around in her bed, something human. At the same time she felt as if someone were shouting in her ear.' It made her feel repulsive, 'like a dog or a devil'. Oddly, Jung made no further comment about this, merely stated the facts, a strange decision which may have had something to do with the fact that he himself knew what it was like to feel repulsive and plagued by the devil. One day Fräulein Spielrein did a drawing relating to her time in Interlaken which showed Dr Heller giving a patient electrical treatment. 'The position is a remarkably sexual one,' Jung noted, leaving it at that. Fräulein Spielrein meanwhile was becoming more and more obsessed with her Herr Doktor.

At first Emma may have thought Fräulein Spielrein, who was neither particularly attractive nor well dressed, was just another of those tragic young women made hysterical by their family circumstances. And normally she would have felt sorry for a young woman so burdened by such a traumatic childhood that a full 'cure' was unlikely. But Fräulein Spielrein was turning out to be trouble. Emma already knew something of Sabina Spielrein. She had even spent time with her in Carl's laboratory where Fräulein Spielrein was helping him collate the word association tests as part of her occupational therapy. 'Patience, calm, and inner goodwill towards the patients' were Bleuler's method 'to remove the basis of hysterical outbursts, and other symptoms, which have merely an attention-seeking nature, by deliberately ignoring them'. Thereafter, as their health and behaviour recovered, the patient needed to be given a purpose in life. In Fräulein Spielrein's case this had started with letting her accompany doctors on ward rounds, because it was known that she wanted to study medicine. Later she was allowed to eat with the assistant doctors and help Herr Doktor Jung collate his word association test results. Emma saw Carl less and less and Fräulein Spielrein saw him more and more. It was all the wrong way round.

Gradually Fräulein Spielrein grew calmer and her behaviour improved, with everyone but the poor female *Wärters* who had charge of her when she was not with her Herr Doktor. 'Suicide attempts to frighten the nurses,' Jung noted; 'running away, giving people scares, transgressing prohibitions'. And if he went away for a few days, perhaps to Schaffhausen with his family, she 'used' his return 'to produce a few scenes': climbing up on the window grille in the corridor, forcing him to pay her an evening visit, or sitting in the doorway of her room in her night-dress wrapped in a blanket, and when no one took any notice, falling into a convulsive fit.

When Carl had to go away for his annual three weeks of mil-itary service Direktor Bleuler took over the Spielrein notes: 'Owing to the absence of the Senior Physician pat.[ient] left almost entirely to herself for the last week, and much worse.' One night at 1 a.m. after making a 'scene', she kicked the nurse who was trying to get her back into bed. The next day she was found clinging to the wall until Bleuler came to talk to her. She hid knives, left 'farewell letters', threatened to starve herself to death, stole the gas worker's ladder from the ward and put benches in the corridor for him to 'jump over'. She demon-strated how she would strangle herself with the curtain cords, and she composed 'songs' about the doctors which she found so funny she could not sing them for hysterical laughing. When Herr Doktor Jung returned there were 'all kinds of pranks': she 'torments the nurse so horribly that she has to be withdrawn'; she 'kicks the stepladder around the corridor' and 'scratches the floor'. She refused to go into town, even though it was part of her rehabilitation programme in preparation for leaving the asylum. She again threatened to starve herself to death. 'She insists that every evening she travels to Mars,' wrote Jung, 'on to which she projects all her contra sexual fantasies.' When she complained of pains in her feet, Jung noted: 'She is afraid of going out and of the future, so she tries to postpone going out as long as possible through the pain in her feet.'

It is not hard to imagine how Emma reacted to all these endless 'pranks' and 'scenes'. Carl had always discussed his work and his cases with her, so she knew what was going on, and she could see for herself how Fräulein Spielrein manipulated events to spend all her time with Carl. So it must have been a relief to Emma when it was decided that Fräulein Spielrein was well enough to leave the Burghölzli and attend as an outpatient instead. But, as Jung had divined, Fräulein Spielrein did not want to leave. The lengthy sessions continued, and she began telling Herr Doktor Jung new and distressing stories about her father: how 'he took her into a special room and ordered her: "lie down"; she implored him not to beat her (he was trying to lift her skirt from behind). Finally he gave in, but he forced her to kneel down and kiss the picture of her grandfather and to swear always to be a good child. After this humiliating scene the boys (her brothers) were waiting outside to greet her.' She admitted to Jung that 'since her fourth year she had experienced sexual arousal'. She had 'orgasmic discharge' and eventually she only had to see or hear one of her brothers being beaten for her to 'want to masturbate'. Even a threat was enough for her to lie on her bed and masturbate. Someone only had to laugh at her, inducing a feeling of humiliation, 'to cause her to have an orgasm'. Jung's notes added: 'during the act, pat. wishes on herself all manner of torments; she pictures these as vividly as possible, in particular being beaten on her bare bottom and, in order to increase her arousal, she imagines that it is taking place in front of a large audience'.

Bleuler wrote to the woman's father in Rostov: 'As her memories of you agitate her greatly, we are of the opinion that Fräulein Spielrein should not write to you directly over the next few months,' and ended the letter: 'Fräulein Spielrein now occupies herself almost daily with scientific reading and she has also commenced practical scientific study in the anatomy laboratory. She wishes to convey her fond greetings. Yours faithfully, the Direktor, Bleuler.'

Sabina Spielrein was finally discharged from the Burghölzli on 1 June 1905, to attend college in preparation for her medical studies at the University of Zürich. 'In the last few weeks distinctly improved,' wrote Jung, 'and increasingly calm. Now listens to lectures conscientiously and with interest.' The treatment could be counted as a success. But if Emma hoped that Fräulein Spielrein would now exit their lives, she was wrong. Soon she was turning up at the Burghölzli for her private outpatient sessions with Jung, fashionably dressed and newly independent, embarked on just the kind of studies Emma herself would have dearly loved to pursue. Worse still, she took to telling anyone who would listen that she was in love with Herr Doktor Jung. And that he loved her. Instead of trying to extricate himself from this highly charged relationship, Jung wrote to Herr Spielrein suggesting he channel the money for his daughter's allowance through himself. Emma was angry, jealous, distressed. But what could she do?

By the autumn Fräulein Spielrein's mother had discovered what was going on. She demanded that Jung write a report on her daughter's treatment for Herr Professor Freud in Vienna, in case her daughter needed to be removed from Zürich for further treatment. Her daughter had already fallen in love with an uncle, and with one of the doctors at the Heller Sanatorium, and she did not want it to happen all over again. She needed to have the report in her possession 'to use if the occasion arises'.

Jung's report, dated 25 September 1905, written on official Burghölzli notepaper, begins with a brief summary of Fräulein Spielrein's history and then continues:

> Masturbation always occurred after she underwent punishment from her father. After a while the beatings were no longer necessary to initiate sexual arousal; it came to be triggered through mere threats and other situations implying violence, such as verbal abuse, threatening hand movements, etc. After a time she could not even look at her father's hands without becoming sexually aroused, or watch him eat without imagining how the food was ejected, and then being thrashed on the buttocks, etc. These associations extended to the younger brother too, who also masturbated frequently from an early age. Threats to the boy or ill-treatment of him aroused her and she had to masturbate whenever she saw him being punished.

It goes on to describe how at the Burghölzli, Fräulein Spielrein 'initially harassed everybody, tormenting the nurses to the limits of their endurance. As the analysis progressed, her condition noticeably improved and she finally revealed herself to be a highly intelligent and talented person of great sensibility. There is a certain callousness and unreasonableness in her character and she lacks any kind of feeling for situation and for external propriety, but much of this must be put down to Russian peculiarities.' Her mother, Jung added, could 'not quite comprehend' why Fräulein Spielrein suffered enormously whenever she met any member of her family, even though she knew about the 'most important part' of her daughter's complex.

Up to this point the report gave an account of a case which Jung had handled professionally and with some success. But the last paragraph reads: 'During treatment the patient had the misfortune to fall in love with me. She raves on to her mother about her love in an ostentatious manner, and a secret perverse enjoyment of her mother's dismay seems to play a not inconsiderable part in this. In view of this situation her mother therefore wishes, if the worst comes to the worst, to place her elsewhere for treatment, with which I am naturally in agreement.'

6

Dreams and Tests

One day during the summer of 1905 Emma sat down at her desk in their apartment on the upper floor of the Burghölzli, and, with Carl at her side, dipped the nib of her pen in the inkwell and began to write. Jung was dictating a dream he had had, and his interpretation of it. Emma was pregnant with their second child. The dream was long and detailed:

> I saw horses being hoisted by thick cables to a great height. One of them, a powerful brown horse which was tied up with straps and was hoisted aloft like a package, struck me particularly. Suddenly the cable broke and the horse crashed to the street. I thought it must be dead. But it immediately leapt up again and galloped away. I noticed that the horse was dragging a heavy log along with it, and I wondered how it could advance so quickly. It was obviously frightened and could easily cause an accident. Then a rider came up on a little horse and rode along slowly in front of the frightened horse, which moderated its pace somewhat. I still feared that the horse might run over the rider, when a cab came along and drove in front of the rider at the same pace, thus bringing the horse to a still slower gait. I then thought now all is well, the danger is over.

Carl went on to dictate a detailed interpretation of the dream, which he would include in his next book, *Über die Psychologie der Dementia Praecox*, which he had been working on since 1903 and would complete in July 1906. The book was, as he explained in the foreword, based on his clinical experiments and researches, so combining 'all the disadvantages of eclecticism, which to many a reader may seem so striking that he will call my work a confession of faith rather than a scientific treatise. *Peu Importe*! The important thing is that I should be able to show the reader how, through psychological investigation, I have been led to certain views which I think will provoke new and fruitful questions concerning the individual psychological basis of *dementia praecox*.' He went on to acknowledge his debt to Herr Professor Bleuler and 'my friend Dr Riklin' and – significantly – to Sigmund Freud in Vienna, a man he had not yet met: 'Even a superficial glance at my work will show how much I am indebted to the brilliant discoveries of Freud.' In Part One he gave a summary of work which was currently being done in the field: by Pierre Janet, Kraepelin, Aschaffenburg, Ziehen, Tschisch, Freusberg, Sommer, Clemens Neisser and Otto Gross – a brilliant doctor who would soon become Jung's patient at the Burghölzli.

The dream came in Part Two. Carl did not admit to it being his own dream but ascribed it to a 'friend', adding that 'the personal and family circumstances of the subject are well known to me'. The analysis took the form of Jung asking the friend what he made of each point of the dream, then offering his own interpretation based on his theory of associations. 'It seemed to him that the horses were being hoisted onto a skyscraper,' Emma wrote, to Carl's dictation; 'X had recently seen in a periodical the picture of a skyscraper being built.' Horses working in the mines were tied with straps in the same manner, his 'friend' recalled, before being lowered down. These 'dizzy heights' brought to mind mountains and the friend admitted that he was a passionate mountain climber, and that:

just about the time of the dream, [he] had had a great desire to make a high ascent and also to travel. But his wife felt very uneasy about it and would not allow him to go alone. She could not accompany him, as she was pregnant. For this reason they had been obliged to give up the idea of a journey to America [skyscrapers], where they had planned to go together. They realised that as soon as there are children in the family it becomes much more difficult to move about and one cannot go everywhere. [Both were very fond of travelling and had travelled a good deal.] Having to give up the trip to America was particularly disagreeable to him, as he had business dealings with that country and always hoped that by a personal visit he would be able to establish new and important connexions. On this hope he had built vague plans for the future, rather lofty and flattering to his ambition.

He went on to point out some further associations: 'to work like a horse' and 'to be in harness' leading to the meaning 'by labour one gets to the top'. On the other hand, being 'hoisted like a package' reminded X of tourists who he had always despised for getting themselves hoisted up to the highest alpine peaks 'like sacks of flour'. Not something he himself ever required, and there was an element of contempt here. So where was the dreamer in all this? Because, as Jung reminded the reader, Freud said the dreamer was usually the chief actor. Surely his 'friend' was the 'powerful brown horse'. But what about the log he was dragging along? Now the 'friend' recalled that 'Log' (or Steam-Roller) had been his nickname when he was younger, on account of his powerful, stocky figure. But, despite the burden, it did not actually hinder the horse. On the other hand, the galloping horse was frightened, and that could easily cause an accident.

Jung had reached an impasse and needed to approach the dream from another angle. Why had the cable broken, causing the horse to fall down to the street? It seemed to associate with an earlier career ambition, which had been thwarted and others preferred over him. 'His disappointment over his failure was so great, he said, that for a moment he almost despaired of his future career. In the dream he thought the horse was dead, but soon saw with satisfaction that it got up again and galloped away.' Jung thought a new section of his 'friend's' dream began here, corresponding to a new period in his life. The friend now remembered something else about the dream: there was another horse, very indistinct, in the background: 'X was dragging the log with someone else, and this person must be his wife, with whom he is harnessed "in the yoke of matrimony",' suggested Jung to X, noting that although the encumbrance might have hindered his progress he was able to gallop away, revealing he could not be tied down. The galloping horse reminded X of a painting by Welti, *Mond Nacht* ('A Moonlit Night'), where galloping horses were shown on the cornice of a building. One of them was a lusty stallion, rearing up. In the same picture there was a married couple lying in bed. The image of the galloping horse, therefore, had led to the very suggestive painting by Welti. 'Here we get a quite unexpected glimpse into the sexual nuance of the dream, where till now we thought we could see only the complex of ambition and careerism,' concluded Jung, adding that this could easily be interpreted as 'X's own impetuous temperament which he feared might involve him in impetuous acts'.

If Emma did not know it before, she knew it now: Carl felt that marriage held him back, but, be warned, nothing would stop him from doing what he wanted to do. That there was a sexual element to it can not have come as a surprise. But did Emma agree with Carl's interpretation of the dream?

'Then a rider came up on a little horse and rode along slowly in front of the frightened horse, which moderated its pace somewhat': Jung first interpreted the 'little horse' as his 'superior', Bleuler. But he had to admit that the horse was 'small and dainty like a rocking horse', which certainly did not fit. Instead, the 'friend' recalled an incident in his childhood when he saw a pregnant woman, and this led once again to his pregnant wife. Another impasse. He turned his attention to the cab which joined the 'little horse' to slow the stallion down, and now the 'friend' saw it was full of 'a whole cartload of children', a fact which had been repressed when he first recounted the dream and only recalled in the telling.

'The meaning of the dream is now perfectly clear,' Jung concluded with a flourish. 'The wife's pregnancy and the problem of too many children impose restraints on the husband. This dream fulfils a wish, since it represents the restraint as already accomplished. Outwardly the dream, like all others, looks meaningless, but even in its top layer it shows clearly enough the hopes and disappointments of an upward-striving career.' He ended enigmatically: 'Inwardly it hides an extremely personal matter which may well have been accompanied by painful feelings.' This is as much as the interpreter of the dream is prepared to reveal about why the stallion was 'frightened'. It was not a word he cared to investigate. Nor the association with the 'little horse' coming to the stallion's rescue, which led to the last thought of the dream to be: 'I then thought now all is well, the danger is over.'

Perhaps Emma realised she was the little horse who came to the frightened stallion's rescue. But she said nothing, or if she did, Carl did not record it. As she wrote and Carl dictated, no one thought it wrong for the analyst to be both subject and interpreter. In later years it became clear that analysis required a practitioner quite separate from the subject to do the interpreting, but not in 1905.

One thing the dream reveals: in real life Carl did not always get his way. When Emma insisted he could not sail off to America or go mountain-climbing because she was pregnant he was forced to agree. After almost three years of marriage Emma was beginning to leave behind the ingénue she was when Carl first met her. In public she remained quiet and reserved, but in private some things were changing. And another thing: for all the years she had been writing down Jung's dictation, writing up reports, helping with research, she was learning first-hand how dreams and associations revealed the hidden desires of the unconscious and the secrets of repression. It was an apprenticeship by default. When the book *Dementia Praecox* was finished, Carl once more dedicated it to Emma, his wife.

Carl sent a copy to *Hochverehrter Herr Professor!* – Professor Freud in Vienna. He had previously sent him a copy of his *Diagnostic Association Studies*, thereby starting the now famous

Emma and Carl at the time of their visit to Vienna. Carl, a penniless doctor, married Emma, one of the richest heiresses in Switzerland, in 1903 – the start of a long and complex marriage.

Sigmund Freud: a photograph taken by his sons and sent as a gift to Carl after the visit to Vienna. Emma had it enlarged and framed for Carl's Christmas present.

Vienna was one of the great cosmopolitan cities of Europe. Only half the inhabitants were German-speaking Austrians; the rest came from the four corners of the Austro-Hungarian Empire, including many Jews, rich and poor.

The two sisters, Emma (*left*) and Marguerite Rauschenbach, in a setting befitting their position in society. They were close but very different: Emma clever and studious, Marguerite outward-going and sporty.

The Haus zum Rosengarten on the far bank of the Rhine was Emma's 'idyllic' childhood home. Large and square, it was the finest in Schaffhausen, with a rose garden by the river, next to the town's washing stand.

The Rhine Falls at Schaffhausen were close to Emma's home. The intrepid could hire a ferryman to row them through the roaring torrent to the rock in the middle, the Swiss flag flying high above.

When Emma was twelve the family moved to Ölberg, a vast Jugendstil mansion above Schaffhausen. By this time Emma's father was blind, so all the plans were produced in braille.

Carl as a boy. 'I had never met such an asocial monster,' recalled a friend. Carl's mother had a second personality and heard voices. His father was plagued with unnamed torments.

Early on Carl sensed problems in his parents' marriage. His mother had to go 'away' occasionally, when Carl was looked after by his father, sleeping in the same room till he went to university, aged eighteen.

Jean Rauschenbach. Emma's father was director of the internationally renowned family business in agricultural machinery until he went blind in 1894, when Emma was twelve.

Emma aged sixteen. Shy and podgy, Emma's 'idyllic' childhood was over. There was a terrible family secret: her father's blindness had been caused by syphilis.

Carl at Basel University. A complicated, ambitious young man, outwardly confident but plagued by deep feelings of inferiority, and with little experience of women.

Carl (*middle row, third left*) dominated discussions at his Zofingia fraternity with his powerful personality, superior intellect and lively humour. His nickname was Walze, 'Steam-Roller' or 'Barrel'.

Emma (*left*) and Marguerite (*centre*) in Canton of Schaffhausen costume. Back from Paris, Emma had turned into a slim, attractive young woman with plenty of admirers.

The Burghölzli was the vast Zürich lunatic asylum where Carl worked from 1900. Under the directorship of Dr Bleuler it was gaining an international reputation for its progressive methods of treatment.

Carl at the Burghölzli. Other staff thought he might have some psychical disturbance because he seemed so alienated. He didn't leave the asylum precincts for his first six months.

One of the men's wards at the Burghölzli. Patients, if not bedridden, were encouraged to take part in the many activities provided under the Bleuler regime.

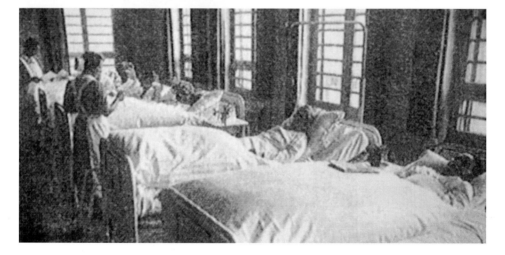

correspondence of 359 letters, not including the missing ones. The letter Freud wrote thanking Carl for his copy of *Dementia Praecox* is lost, but it is clear from Jung's reply that Freud offered a somewhat different interpretation of the horses dream, suggesting it related to 'the failure in a rich marriage'. Freud had rumbled that the dreamer was not Jung's 'friend' but Jung himself.

'You have put your finger on the weak points of my dream analysis,' Jung wrote in reply to Professor Freud on 29 December 1906:

> I do in fact know the dream material and the dream thoughts much better than I have said. I know the dreamer intimately: he is myself. The 'failure of a rich marriage' refers to something essential that is undoubtedly contained in the dream, though not in the way you think. My wife is rich. For various reasons I was turned down when I first proposed; later I was accepted, and I married. I am happy with my wife in every way (not merely from optimism), though of course this does nothing to prevent such dreams. So there has been no sexual failure, more likely a social one. The rationalistic explanation, 'sexual restraint' is, as I have said, merely a convenient screen pushed into the foreground and hiding an illegitimate sexual wish that had better not see the light of day.

Jung then offered another interpretation of the 'little rider' as the wish for a son, adding 'we have two little girls' – baby Gretli having been born on 8 February 1906.

It would be surprising if this explanation convinced Freud. Jung also insisted, presumably with reference to another of

Freud's suggestions in the lost letter: 'I have been unable to discover an infantile root anywhere.' Clever Freud had put his finger on all sorts of repressed associations, but neither Jung nor Freud appear to have considered the possibility that 'the little horse' represented Emma, coming to the rescue. 'The analysis and use of one's own dreams is a ticklish business at best,' Jung admitted to Freud; 'one succumbs again and again to the inhibitions emanating from the dream no matter how objective one believes oneself to be.'

Freud wrote to his 'esteemed colleague' on 30 December, disingenuously giving him an example of a case of a female patient of his who had suffered from depression. 'She is very much in love with her husband (he is an actor) but has been totally anaesthetic in sexual intercourse. The patient adds that it has never occurred to her to blame her husband for her lack of satisfaction, that she is convinced it is her fault.' He worded it as though he was asking for Jung's opinion in a professional matter, ending: 'Forgive me for taking up your time.' Apparently Freud did not believe Jung's 'no sexual failure, more likely a social one'. But then Freud knew nothing of Carl's complex Personality No. 2.

On New Year's Day 1907 Professor Freud wrote again. He had meant his previous letter to be longer but: 'I broke it off partly for incidental reasons and partly because my guess, confirmed by you, as to the identity of the dreamer, bade me hold my peace. I merely thought you might have gone so far as to stress the interpretation log = penis and the "alternative" gallop, horse/career, without giving yourself away.' He contented himself with pointing out that Jung's version of the wish-fulfilment, to have a son, was hardly credible. After that he concentrated on their hoped-for future partnership. 'The "leading lights" of psychiatry really don't amount to much,' he told him, 'the future belongs to us and our views, and the younger men – everywhere most likely – side actively with us. I see this in Vienna, where, as you know, I am systematically ignored by my colleagues and periodically annihilated by some hack, but where my lectures nevertheless draw forty attentive listeners, coming from every faculty.' He hoped Jung would come to Vienna soon, 'before you go to America (it is nearer).' Still, he could not end the letter without some further reference to his usual topic: 'I should like to observe that you omit a factor to which, I am aware, I attribute far more importance than you do at the present time; as you know, I am referring to xxx sexuality.' Three crosses meant danger. They used to be chalked on the inside of doors in peasant houses to ward it off.

Jung did not answer for a week, and when he did he apologised for not writing sooner. Presumably Freud's letter had given him plenty to think about and he took his time. 'Afterwards I was rather embarrassed at having played hide and seek with my dream,' he admitted, adding: 'There are special reasons why I did not bring in the interpretation log = penis, the chief of which was that I was not in a position to present my dream impersonally: my wife wrote the whole description (!!).' Perhaps Emma was doing more than merely taking down Carl's dictation by offering interpretations of her own. After this admission his letter quickly reverted to professional talk and the hope that he might visit Freud in Vienna during their forthcoming spring holiday.

Emma was meanwhile beginning to take a more active role at the Burghölzli asylum. In 1907 Dr Abraham Brill (known as A.

A. Brill) arrived from America to join the staff, replacing Karl Abraham who returned to Germany. Brill's account of the way Bleuler's 'active community' involved everyone, each to their own capacity, is full of the atmosphere of those heady times. He points out that the community included the wives:

> In the hospital the spirit of Freud hovered over everything. Our conversation at meals was frequently punctuated by the word 'complex', the special meaning of which was created at the time. No one could make a slip of any kind without immediately being called upon to evoke free associations to explain it. It did not matter that women were present – wives and female voluntary interns – who might have curbed the frankness usually produced by free associations. The women were just as keen to discover the concealed mechanisms as their husbands. There was also a Psychoanalytic Circle which met every month. Some of those who attended were far from agreeing with our views; but despite Jung's occasional impulsive intolerance, the meetings were very fruitful and successful in disseminating Freud's theories.

Whatever else Emma's life in Schaffhausen might have provided, it could never have competed with the fevered atmosphere at the Burghölzli asylum in those early years. 'Jung brooked no disagreement with Freud's views,' Brill added, 'impulsive and bright, he refused to see the other side. Anyone who dared doubt what was certainly then new and revolutionary immediately aroused his anger.'

Auguste Forel, the former Herr Direktor still wandering around the corridors of the Burghölzli, was distinctly fed up about it. 'This Freud cult disgusts me,' he wrote to his friend Ludwig Frank. 'I leave open the question if the famous discovery of Freud is really his and doesn't rather belong to Bleuler but it is certain that in Vienna, where people aren't prudish, Freud has a very bad reputation which is not unfounded . . . It appears to me as if Bleuler is no longer the Director of the Burghölzli, but Jung, and I am sorry.'

Jung was certainly getting into his stride. 'It is amusing to see how the female outpatients go about diagnosing each other's erotic complexes although they have no insight into their own,' he wrote to Professor Freud in June, quite forgetting the trouble he had with his own. Ten days later he went on a trip to Paris and London, furthering his career and leaving Emma back at the Burghölzli with the two little girls. Once he was back home he wrote to Freud about 'a German-American woman who made a pleasant impression on me – a Mrs St, aged about 35', whom he had met in Paris. They were at a party, chatting about landscapes, when 'Mrs St' declined the offer of black coffee saying she could not tolerate even one mouthful. Impetuous Carl plunged in with an unasked-for interpretation: this was a nervous symptom, he suggested, and would only apply when she was at home; elsewhere she would tolerate the coffee better. 'Scarcely had this unfortunate phrase left my mouth than I felt enormously embarrassed,' he admitted to Freud, 'but rapidly discovered – luckily – it had "slipped by" her.'

In the same letter Jung wrote: 'My wife, who knows a thing or two, said recently: "I am going to write a psychotherapeutic handbook for gentlemen."' Having attended all the meetings, with their talk of erotic complexes, repressions and slips of the tongue, and participated in research and the word association tests, Emma, on the eve of their first visit to Freud in Vienna, was learning fast.

At 19 Berggasse that March, 1907, Freud watched Emma Jung: the reserved but friendly way she behaved at that midday Sunday meal – intelligent, interested, talking to Martha and Minna and the children at one end of the table whilst her husband talked his head off down the other. So when Freud received Jung's letter in July he probably was not surprised to hear that Emma Jung 'knew a thing or two' about psycho-analysis.

For Jung the Vienna visit had been thrilling, talking with Freud late into the night. But the after-effects were traumatic. Too much had been stirred up with all the talk and the flattering admiration, and above all Freud's implacable insistence that sexuality was the root cause of all neurosis, which Jung struggled to accept. 'But don't you think that a number of borderline phenomena might be considered more appropriately in terms of the other basic drive, *hunger*; for instance, eating, sucking (predominantly hunger), kissing (predominantly sexuality)?' Jung had written in October 1906, even before meeting Freud in person.

Growing up in a poor, isolated Swiss peasant village he knew all about hunger. And not only hunger. 'I had grown up in the country, among peasants, and what I was unable to learn in the stables I found out from the Rabelaisian wit and the untram-melled fantasies of our peasant folklore,' Jung wrote years later. 'Incest and perversions were no remarkable novelties to me, and did not call for any special explanation. That cabbages thrive in dung was something I had always taken for granted.' The trouble was, he himself had been sexually abused as a boy, so, ironically, he was an example of Freud's theory. And with all this talk about sex he was no longer able to repress it. It was another two years before he could admit to Freud that the Vienna visit had culminated in an obsession with the lady in Abbazia. 'As I have indicated before, my first visit to Vienna had a very long unconscious aftermath, first the compulsive infatuation in Abbazia, then the Jewess popped up in another form, in the shape of my patient. Now of course the whole bag of tricks lies there quite clearly before my eyes.'

In fact Freud had recently shifted his views on childhood sexuality, deciding that some patients who claimed to have been sexually abused were, in fact, fantasising. Karl Abraham agreed with Freud: one of the reasons there appeared to be so many cases of child sexual abuse was that 'the trauma [was] wished for in the unconscious, so we have here a glimpse of a form of infantile sexuality'. The child was fantasising, or, if a real act, was to some extent a willing participant. Why, otherwise, did the child not call for help? Or run away? Thence the feeling of guilt in the child, suggested Abraham. But Jung knew he was not fantasising. And he knew that in isolated villages, where families were already too large, there were plenty of cases of incest. And he knew from personal experience that abused children do not always run away, or tell anyone about it. They repress it. 'You will doubtless have drawn your own conclusions from the prolongation of my reaction-time,' he told Freud. 'Up till now I have had a strong resistance to writing because until recently the complexes aroused in Vienna were still in an uproar. Only now have things settled down a bit, so that I hope to be able to write you a more or less sensible letter.'

In August, Jung was again apologising to Freud for his 'long silence'. He had been away on his annual three weeks' army military service: 'at it from 5 in the morning till 8 at night' and then, dog-tired, back to the Burghölzli where he was over-worked, with never enough time to do his own research. 'Often I want to give up in sheer despair,' he wrote. By September he was at a congress in Amsterdam defending Freud's reputation. He spoke without his usual clarity and confidence and had to be called off the podium for overrunning, 'refusing to obey the chairman's repeated signal to finish. Ultimately he was compelled to, whereupon with a flushed angry face he strode out of the room,' as Ernest Jones, a young Welshman and a keen disciple of Freud, later described. When Jones had first met Jung, during a visit to the Burghölzli to observe the Herr Doktor's methods, he noted his 'formidable presence', but now, seeing this lamen-table performance, he was not so sure. Jung knew it. Writing to Freud afterwards, still smarting from the humiliation, he described the crowd as 'ghastly, reeking of vanity' and the discussions 'a morass of nonsense and stupidity!'

By the end of September, Carl was in bed with severe gastro-enteritis, requiring Emma's nursing and stabilising. A new voluntary assistant at the Burghölzli, Max Eitingon, was getting under his skin but it took Jung a while to admit why. 'I consider Eitingon a totally impotent gasbag,' he wrote to Freud. 'Scarcely has this uncharitable judgement left my lips than it occurs to me that I envy him his uninhibited abreaction to the polygamous instinct.' Eitingon had many affairs quite openly, and Jung envied him.

Emma and Carl spent Christmas and the New Year, 1908, with Emma's mother at Ölberg, whence Emma sent Freud and his family a New Year's greeting. Though she had not spent a great deal of time with Freud in Vienna, it had been enough to appreciate his charm and the cleverness with which he handled her ebullient husband. This Christmas, Emma had a surprise present for Carl: an enlarged and beautifully framed photograph of the Herr Professor. Jung had been asking Freud to fulfil 'a long-cherished and constantly repressed wish' for a photograph, and Freud had finally sent one – taken informally by 'my boys', using a roughly hung curtain as backdrop. On 25 January, Jung wrote to thank him in terms which might have surprised the Herr Professor: 'I have a sin to confess: I have had your photograph enlarged. It looks marvellous.' A sin? What Emma thought of Carl's claim to have done it himself is not recorded. Freud forbore to comment on the photograph in his reply, telling Jung about his health instead: he had recovered from his bout of influenza which had lasted long enough: 'For six long weeks the frog was ill/But now he's smoking with a will,' he wrote, quoting his favourite Wilhelm Busch, and taking no notice of the doctor's warnings about what might happen to him if he did not give up smoking those cigars.

Carl had also been ill again, but by the beginning of January he was already back at the Burghölzli leaving Emma and the girls behind in Schaffhausen. 'At the moment I am treating another case of severe hysteria with twilight states. It's going well,' he wrote to Freud with evident relish. Originally Hans Maier, the first assistant, was meant to conduct the analysis with the patient, but the woman resisted, having already 'set her cap' at Herr Doktor Jung. Doctors and nurses clustered round when the patient was in one of her twilight states, full of wonderment at the dramatic beauty of her utterances. 'At present she is expecting a visit from her lover, but is afflicted with a ructus [burping]. She is always standing at the window looking out to see if he is coming.' Two weeks later Jung was wowing a mesmerised public in the auditorium of the Zürich city hall. Up one minute, down the next, was the pattern: by February he was ill again, this time with 'a beastly attack of influenza'. He felt run down and listless and Emma had to steady him once more. His scientific work had come to a total stop. He needed a holiday.

'All sorts of psychogenic complications insinuated themselves into my influenza, and this has had a bad effect on my convalescence,' wailed Jung. 'First of all a complex connected with my family played the very devil with me, then I got disheartened by the negotiations over the journal.' The journal – *Jahrbuch für psychoanalytische und psychopathologische Forschungen* – was close to Freud's heart, the means by which Freudian psychoanalysis could reach an international public, and he was determined that Jung should be the editor. As to the 'complex connected with my family', we can only guess what it was about: maybe his age-old feelings of social inferiority, his 'vaulting ambition' which led him again and again to overwork and exhaustion. But the most likely reason was quarrels with Emma over his continued involvement with Fräulein Spielrein. Emma probably knew it was not a fully sexual relationship, but it was a highly charged emotional one – much more than the usual 'infatuations' – and Carl seemed quite unable or unwilling to break it off.

Emma was at her wits' end, mortified and jealous. At some point she threatened Carl with divorce, either before or after their visit to Vienna, but certainly once she realised Fräulein

Spielrein was still making trouble, spreading rumours. Emma knew Carl. She knew that the last thing on earth he wanted was divorce. Now she summoned all her courage to have it out with him. Her own childhood had not prepared her for a husband with 'polygamous instincts'. Her father had been a good husband to Bertha, caught out by a transgression on a business trip to Budapest. It is not known how Emma handled the showdown with Carl. She was not used to making scenes. It was not in her nature and it was humiliating. But scenes she made. 'It seems a very long time since I last wrote to you,' Jung wrote to Freud from the Grand Hotel Bellevue, Buvento, on 11 April 1908. 'All sorts of things have played the devil with me, for instance a hideous bout of influenza that left me so debilitated that I had to take thermal baths at Baden. Now I'm setting about recovering as thoroughly as possible on Lago Maggiore.' He would write again once he was back home. 'At present I am too dissociated.'

'Divorce/force' was one of the word associations in the long series of tests Jung underwent for Ludwig Binswanger, spread across several weeks. Binswanger's family ran the Bellevue Sanatorium at Kreuzlingen and he came to the Burghölzli asylum like so many others, attracted by its international reputation and that of Herr Doktor Jung. Jung, always generous to students with his time, agreed to be Binswanger's supervisor and offered himself as one of the 'educated subjects' for the tests. This came a month or two before the Jungs' second visit to Vienna, otherwise Carl might well have refused, because the word associations turned out to be much more revealing and unsettling than he had anticipated.

Binswanger's dissertation was titled 'On the Psychogalvanic Phenomenon in Association Experiments'. He used twenty-three subjects in all, taken from the usual source: educated and uneducated, some patients, some doctors, some *Wärters*, male and female. He made use of the galvanometer again. 'I will begin with an experiment on a subject remarkable for penetrating and reliable self-analysis – a married doctor well acquainted with the association experiment,' he wrote.

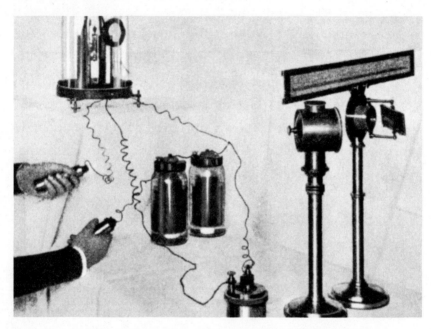

The galvanonmeter – a scientific technique.

Using the Rauschenbach stopwatch he noted that the subject's mean time for a reaction was 1.8 seconds. 'Head/band' it started, followed by 'blue/sea'. But then came 'wall/star', an odd association which Binswanger asked the subject to explain. Apparently the subject meant to say 'stone' but deviated from it. There followed a long explanation of the deviation, none of which mentioned the fact that a stone in the garden wall of the parsonage where Jung lived as a child had played such a vital part in his young life, when he had tried to decide whether he was the stone or the stone was himself.

Then Binswanger, seeing Jung was too well prepared, changed the stimulus word from 'angel' to 'devil'. That took Carl by surprise. The devil played a critical part in the secret life of the 'other Carl', as did religion in general, albeit in an unconventional way.

'My veneration for you has something of a "religious" crush,' Jung had confessed to Freud in the letter in which he also disclosed he had been sexually assaulted by someone he had 'worshipped' as a boy. Five days later, waiting for Freud's reply which was unusually long in coming, he wrote again: 'I am suffering all the agonies of a patient in analysis, riddling myself with every conceivable fear about the possible consequences of my confession.' Freud's reply is missing but it evidently managed to calm Jung about his 'confession'. It worked wonders for him, he said, and you can hear the relief: 'You are absolutely right to extol humour as the only decent reaction to the inevitable. This was also my principle until the repressed material got the better of me, luckily only at odd moments. My old religiosity had secretly found in you a compensating factor which I had to come to terms with eventually, and I was able to do so only by telling you about it. In this way I hoped to prevent it from interfering with my behaviour in general.' Freud answered this one in typical fashion, combining good advice with humour: 'A transference on a religious basis would strike me as most disastrous; it could end only in apostasy, thanks to the universal human tendency to keep making new prints out of the clichés we bear within us. I shall do my best to show you that I am unfit to be an object of worship. You probably think I have already begun.'

When Binswanger got to 'will/strive' there was a long deviation because 'the subject has a strongly pronounced striving for knowledge, work and recognition. A strong complex.' This was followed by 'friendly/hateful', to which Binswanger comments: 'Wanted to say *hate*. Constellated by a dream of the previous night which referred to an old complex not recently revised and which will be dealt with later. So far there has been a rapid rise of the whole curve.' To 'thorn' Jung responded 'flesh'. 'It at once occurs to him that *thorn* signifies a sexual symbol and has replaced the similar *stalk*. During the reaction he is thinking that [Saint] Paul says of himself he has a thorn in his flesh.'

'Dance/joyous' was accompanied by a big increase in the curve on the graph, because, said Jung, he was annoyed that he had to accompany his wife to a ball the next day. 'Sea/lake' came next: he had a journey to England planned. 'Proud/eminent' elicited the comment: 'he is often reproached by his family with being too proud! He has, in general, a strongly developed conscience.' It was followed by a rise in the curve for 'oil': '*öl/berg*', the name of Emma's family estate in Schaffhausen. 'Threaten/hit' aroused the same complex as an earlier association 'on account of which he was for a time threatened'. 'Rich/poor' caused the biggest deviation so far: 'Money complex plays a large part in this subject and is closely connected with his love of travelling, his salary at the asylum, and his future way of living.' The same applied to 'money/have'. With 'play/cards' the subject thought of his wife and children. Then came 'repent/spent' and another rise of the curve denoting 'a rising emotional tone' and referring back to the same old 'not gone into' complex.

A footnote comments: 'For personal reasons this complex was not analysed more fully.'

From Association 51 onwards 'the subject admits he experienced distinct excitation'. This followed 'belly/lie' and 'wrong/right', so it was back to the same old complex again. It is at this point that 'divorce/force' was introduced, followed by 'child/have' where Binswanger, alerted to Jung's deviations, comments: 'He thinks that his youngest child is slightly unwell. But there seems to be some stronger complex behind this apparently "innocent" reaction.' He goes on to quote Freud's *Interpretation of Dreams*, that superficial reactions often hide 'a real and deeper link' – a complex and the resistance to it. 'It is as if the barricade, e.g. an inundation in the mountains, had made the large and best road impassable,' wrote Freud, 'transport then becomes restricted to the inconvenient and steep bridle-paths, which are otherwise only used by the hunter.'

In Jung's case these steep paths often led to his father, the pastor, and to his own 'religiosity', Binswanger noted, adding that 'the subject was very religious in childhood'.

The climax came in a series of tests conducted about a month later during which Binswanger used further methods of distraction to penetrate Jung's resistances: a metronome set fast at ninety-four beats a minute, electrodes on the soles of the feet, and a beam of light shone in the eyes. The subject is described as the married doctor again, but from Binswanger's point of view we are told: 'He is also reserved in ordinary conversation. This reserve is a defence measure against his extreme sensibility, but it is not so much inherited as acquired from various events which have acted upon him as boy and youth, which even today are still to some extent active . . . The bright, careless boy early became a very serious man who has developed, as compensation, a most unusual zeal for duty, and anxiety about work.' Apparently the subject was now 'in a state of active opposition to the analysis'.

It had begun happily enough with 'head/beautiful', apparently an association to the subject's wife. For 'friendly/beloved' the subject commented that 'beloved' was always a word that stood for his wife, and he thought of his wife again with 'dance/much' and with 'housekeeping/wife', followed by 'wicked/good', both 'strongly emotionally toned'. Likewise 'child/mother'. To 'people/state' he commented: 'the State makes such demands upon him that he has little time to devote to his wife', the state in this case being the Canton of Zürich. Other insights into the subject's preoccupations included 'ride/well': 'During his military service he would like to have learnt to ride well, but an illness prevented his riding.' Since Emma was a fine horsewoman perhaps it is not surprising that this comment is followed by: 'He admits that this worried him afterwards. We see that everything that hinders him in his ambition provokes vivid emotional expressions.' Glass/drink 'reminds him of the teetotal movement, of which he is a fervid partisan'.

But there is an undercurrent running along which Jung was trying hard to avoid. Binswanger noted that the subject's whole upper body and legs were moving about. Certain sounds agitated him, he noted, especially: 'Shortly before the experiment the subject had been informed that a former female patient, whose name begins with this sound, had slandered him.' The sound was an 'S': Sabina Spielrein. At the end of the test Binswanger made Jung calm down and rest quietly, whereupon he fell fast asleep.

7

A Home of Their Own

Apart from threatening divorce, Emma now insisted on moving out of the Burghölzli and having a house of her own. The rumours flying about the place were too hurtful. Apparently Herr Doktor Jung was on the verge of leaving her to marry Sabina Spielrein. Emma knew it was not true and that most of the rumours emanated from Fräulein Spielrein herself. But no smoke without fire. It was mortifying, knowing people were talking about you, feeling sorry for you, wondering how you put up with it. People whispered about her wealth too, and the fact that Carl had not had a penny when they married – the inference being obvious. She had had enough. She wanted to have a home of their own where they could lead a proper, private family life.

It didn't take much insistence on Emma's part. All his life Jung had longed for such a family house, dreamed of it, had visions of it. Money not being an object, they decided to build their own. They purchased a plot of land at Kusnacht, a village on the right bank of the Lake of Zürich some fifteen kilometres from Zürich itself, which could be reached by paddle steamer or the local railway. There was a country road along the lake, which would later become the busy Seestrasse, but in 1908 the village was just a small farming community with orchards, meadows for the cows, vineyards rising on the slopes and heavy beds of reeds all along the waterfront. There was an orphanage, a poorhouse, a church, a schoolhouse, and a few scattered villas which had been built in the 1890s when the railway network reached Kusnacht, bringing the wealthy of Zürich who were looking for a pleasant spot to build their homes away from the increasing noise and bustle of city life. The plot Carl and Emma bought was right on the lake, next to the poorhouse: 5,000 square metres at eleven francs a metre – a great deal of money in the days when the average office worker was lucky to earn 800 francs a year. Their neighbour was a farmer called Hermann Stahli, who lived in a typical vintner's farmhouse, solid and rectangular, dating from the seventeenth century, with a steeply pitched saddle roof and elevated ground floor, a wine cellar, barns and stables. Further along the lake the reeds had been cleared for a small communal bathing station.

Emma knew how much such a house would mean for Carl and no doubt she hoped it would have a steadying, grounding effect on him. But she probably did not realise how profoundly significant it was to Carl's Personality No. 2, a personality she lived with day by day, without understanding or even completely recognising it.

Carl's Personality No. 2 had already seen his 'house' when he was a twelve-year-old schoolboy in Basel, suffering from depression, and 'No. 1 wanted to free himself from the pressure or melancholy of No. 2. It was not No. 2 who was depressed, but No. 1 when he remembered No. 2.' That day a wild wind had lashed the Rhine and during his long walk home from school he had seen a ship with a great mainsail running up river before the storm. The sight thrilled Carl and set off his vivid imagination. He saw water all around, like the sea or a lake, with a rock rising up out of it. 'And on the rock stood a well-fortified castle with a tall keep, a watch-tower,' he wrote. 'This was my house. In it there were no fine halls or any signs of magnificence. The rooms were simple, panelled, and rather small. There was an uncommonly attractive library where you could find everything worth knowing. There was also a collection of weapons, and the bastions were mounted with heavy cannon.' The vision was so vivid he indulged in it for several months, making his long walk home so short he would suddenly find himself in his village. There was a two-masted schooner in the harbour because ever since Carl could remember he had longed for a sailing boat. But 'the nerve centre' and main significance of it all was 'the secret of the keep, which I alone knew'. It was hidden inside the tower: a copper column as thick as a man's arm with 'a certain inconceivable something' down in the cellar where there was a laboratory in which Carl made gold. The castle came from another era and another world where there would be 'no *Gymnasium*, no long walk to school, and I would be grown up and be able to arrange my life as I wished'. From then on he had started building miniature castles and fortifications in the garden of the parsonage using stones and mud, and did so for the rest of his life, whenever he needed to calm his mind. But it was the secret of the keep which held the key, mirroring the phallus dream he'd had when he was four years old.

Within a month of buying the plot Carl had already drawn a rough sketch of the property and worked out some of the details of the house. The sketch is done in pencil on a piece of brown paper, indicating the shape of the property, the country road behind the house and the lake in front. The boathouse for a future sailing boat is already there, the word 'boat' written boldly in blue, as is the word 'lake', with 'natural bank with reeds' also marked. The house itself is to the left and rectangular with a tower at the front entrance and a veranda at the side, surrounded by a garden with a pavilion by the lake in the style of the local vintners' huts. Orchard, vines, garden wall and small gate are all included, as they soon would be in reality. There are three plans showing the three floors, all small rooms as in his schoolboy fantasy, except for the main room on the ground floor which was as large as a medieval hall. The drawings were enclosed in a letter sent to Ernst Fiechter, an architect and cousin of Carl's. He may also have included some drawings of houses he had done earlier, all looking like small castles, always with towers, built of the local Bollenstein stone, with crenellations, high saddle roofs, rusticated ground floors, small grilled windows to keep the house safe from the outside world, and walled gardens – like buildings from another era, medieval or Renaissance, and quite unlike the large villas built on the shores of the lake by wealthy Zürich burghers at the beginning of the twentieth century.

'You can get an approximate idea of the external character from my ever-so imperfect sketch,' he wrote in the accompanying letter to Fiechter. 'Rough ashlar below; plasterwork and splatter-dash above. It is to have stepped gables with small steps ... All windows on the raised ground floor are to have grates (*sic volo, sic jubeo*, thus I will, thus I command)' and so on for a good two pages, the pronoun always 'I', never 'we'. The hall should have simple brown wainscoting to head height and a large fireplace of red brick. The ceiling wooden, the floor red-tiled. This was the *Stube*, the main family room, in white stucco, where the tapestries Emma brought from Schaffhausen would be hung, with decoration on the doors and ceiling in the rococo style, and the floor to be raised by one step. There should be one other large room, on the upper floor: the library, separated from Carl's 'workroom' by double doors. All the small rooms to be wallpapered quite simply, the floors simple parquet, the windows with small leaded panes. Downstairs the windows high and narrow, the hall ones of a Renaissance character, again leaded, some heraldic. Kitchen and bathroom: all simple and functional, with an additional bathroom for the servants in the basement. The veranda: white and rococo. Plus a wooden boathouse up to four metres in length. The small yard by the lake to be surrounded by a wall and symmetrically laid out with country flowers; the rest of the garden like a meadow with apple and pear trees. There would be a simple balcony with iron railings for the library on the second floor. On and on it went, down to the last detail. He signed off with the hope of hearing from Fiechter soon, with an approximate estimate, reminding him that the furnishings of the house 'are to be as simple as possible. (With the exception of the reception room and hall.)' The words 'small' and 'simple' pepper the letter without the slightest regard for reality.

Emma was the one concerned with reality. 'Much of it struck us as laboured and strange, not beautiful,' she wrote to her mother about an exhibition of modern villas they were visiting at Darmstadt on one of their holidays without the children. 'Moreover, the houses seemed to be somewhat light in construction and in some cases already looked rather *passé*. The gardens by contrast were charming; it would be nice if Uncle

Mertens could have a look at them sometime.' Uncle Evariste Mertens was their next-door neighbour at Ölberg and the one who had designed the gardens for her parents, even the gardens along the Rhine for the Haus zum Rosengarten, all those years ago. Unlike Carl's letter, Emma's is full of 'us' and it sounds as though the two of them, Emma and Carl, walking around the exhibition arm in arm, were in complete agreement about what they considered 'beautiful': nothing too modern or self-conscious which would soon look *passé*, and something solid and well built. Later, on their next spring holiday on Lake Maggiore, Emma again wrote to her mother: 'On our morning walk we saw many fine villas with splendid southern gardens and have observed with satisfaction that there are still many people who spoil themselves even more than we do.' Emma, modest by temperament, was feeling uneasy as the budget for the house rose higher and higher.

Bertha Rauschenbach was taking a keen interest in the project. 'View received with thanks,' Jung wrote to Fiechter on 14 March 1908, shortly before he and Emma set off on the holiday. 'My mother-in-law will certainly take pleasure in a copy. She was always delighted with your plans.' If Ernst Fiechter was alarmed at Carl's overly detailed plans he did not show it; he just went ahead and produced a charming set of sketches and drawings of houses firmly in the local Lake of Zürich style. Emma must have had a hand in this – they certainly do not look the least bit medieval. They do not even include a tower. At which point Carl put his foot down. How could he live in a house without a tower? He had the boathouse, he had the lake, he had the large hall, he had the veranda and the garden pavilion, but where was his tower?

It appeared in the next set of drawings, a month later, at the front entrance: a large double tower. Two weeks later came the final design: with a single tower, functioning as a stairwell. It came with an update of the budget. The 'garden hall': 4,600 francs; 'courtyard, driveway, and enclosure toward street': 2,500 francs; 'garden by the lake': 4,000 francs; 'boathouse with landing stage': 1,600 francs.

Jung replied that 'we' would like 'to dispense with the garden hall for the moment', but 'we' would like the garden pavilion enlarged instead 'so that a little table and a few chairs can be placed inside'. The 'we' was Emma insisting on some savings: 4,600 francs for a garden hall was just too much. Thereafter Carl reverted to the more typical first person. 'After that, my desire is to have an iron harmonica shutter on the three-part window in the study,' followed by detailed instructions about all the other shutters on all the other windows. 'The little garden gate can stay right next to the house, but best to run the garden wall directly from the corner of the house to the neighbouring property. Likewise, I would like a wall from the house to the lake running along the neighbour's property to the north, thus creating a proper termination there (see sketch 3). A gateway at the house large enough that a carriage can be driven in comfortably. As for the rest, nothing has changed.' Carl, so progressive in his treatment of patients, was in many ways a typical husband of the times: as head of house he expected to take all the decisions. And Emma did as the books on how to be a good wife advised: she submitted to her husband, but learnt to quietly circumvent him when he wasn't looking.

Imagine the hours and hours Emma and Carl spent poring over plans – the discussions, the excitement. Somehow Carl managed to fit it all in: his job at the Burghölzli, writing academic papers in the evenings, the constant correspondence with Herr Professor Freud, and the work involved in founding the journal. Plus 'The Psychology of *Dementia Praecox*', 'On Disturbances in Reproduction in Association Experiments', 'Association, Dream, and Hysterical Symptoms', 'Further Investigations on the Galvanic Phenomenon and Respirations in Normal and Insane Individuals', 'The Content of the Psychoses', 'The Freudian Theory of Hysteria', 'The Analysis of Dreams' – all written during this period, at a time when psychoanalysis was taking its first shaky steps. On top of which came his lectures. The one he gave on 16 January 1908 on 'The Content of the Psychoses' was so oversubscribed they had to move it from the university's main lecture hall to the auditorium in the city hall.

The plans for the house continued apace. By 1 May 1908 Carl was able to write to Fiechter: 'We are in complete agreement with the design and no longer wish to anticipate fate any further. The solution is, as I have said, very satisfying! Even the veranda promises to be most handsome.' The 'solution' is telling: less a medieval castle, more a building in the Zürich style, the *Heimatstil*, fitting in nicely with its Kusnacht neighbours, but with a tower at the front entrance: a happy compromise. Emma had managed to hold her own. There only remained the small matter of the builder's shockingly high estimate: 156,000 Swiss francs. 'I believe a series of cuts still have to be made,' Carl wrote to Fiechter a few days later – 'Ugh!' But none were implemented.

By 4 July 1908 Fiechter wrote to say that work would begin the following Monday.

Now Carl turned his attention to words to adorn the building. The entrance was in the neo-baroque style with a broken pediment, decorated with garlands and fruit above, laurel wreath below, encircling an inscription. 'What will the inscription and decoration over the entrance be like?' he wondered in a letter to Fiechter on 18 September, before giving him the answer: 'I would like to propose the following, deferring to you: *Carol. Gust. Jung et uxor ejus Emma Rauschenbach hanc villam ridenti in loco otioso erigere iusserunt anno domini MCMVIII.* [Carl Gustav Jung and his wife Emma Rauschenbach built this house in a cheerful, tranquil place in 1908] . . . This way the inscription is quite long, as is befitting the style. The house already looks very good.' And beneath it, on the lintel, in larger Roman lettering: '*Vocatus Atque Non Vocatus Deus Aderit*': Summoned or Unsummoned, God Will Be Present. It was the Oracle of Delphi, which Jung knew from a volume of classical sayings and proverbs published by Erasmus of Rotterdam, which he had bought when he was nineteen, presumably adding to his eventual 3,000-franc debt.

By March 1909 the building of the house was finished. One last message to posterity was added, in the ball beneath the weathervane which crowns the stairwell tower. It contains a lead box with a page from a newspaper dated 12 March 1909 and a card in Jung's hand, written on both sides in blue, a colour he reserved for the most important matters. On one side, all in English: '1908–1909. Carl Gustav Jung, med Dr, and his wife Emma had this house built in 1908', followed by a quote in English from Shakespeare's *Twelfth Night*: 'Fate, show thy force – ourselves we do not owe; what is decreed must be, and be this so.' On the other side of the card, in German: 'You who will read this in a later age will perhaps not know my name, and my fates are unknown to you. When I wrote this, I was in the middle of life, perhaps as you are, and did not know the fates of the future. You and I will both succumb to transience. I will simply precede you.'

The master plumber, August Keller of Kusnacht, who installed the lead box, was not impressed. 'Opening this will not be worth the effort. The plumber Aug. Keller. Anno 1909' was inscribed on the outside. Inside he added more, dated 23 March 1909: 'To the openers! The master plumber to whom the execution of this metalwork was entrusted has also taken the liberty of enclosing a few lines in haste. My wish would have been for the builder Dr Jung-Rauschenbach, doctor at Burghölzli in Zürich, to have placed important documents and coins inside this ball, but it seemed to me he prefers to use the coins for something else . . .' and ended: 'I send my best compliments to the openers of this lead box, I will have long since ceased to be among the mortal, to any descendants I may have, I send regards. Aug Keller Jr. master plumber, born 1881. Citizen of Kusnacht.' Both messages were discovered in October 2005 when the stairwell tower was being repaired.

The finishing touch was the weathervane high above, crowning the tower, turned this way and that by the winds which blew down from the Alps or across the Lake of Zürich with the early morning mists, shining golden in the summers, covered in icicles and snow in the winters. 'CJ ER – AEDIFIC – Ao 1908' it announced to the world, as if the world did not already know it.

* * *

Before CJ and ER moved to the house there was another trip for the Jungs to see Freud and his family in Vienna. Again the girls were left in the care of the maid and Carl's mother and sister, who had already moved out of Zürich and into an apartment in one of the Kusnacht houses further along the lake. This time Jung wrote to Freud to say they preferred to stay at the Hotel Regina, the Grand Hotel being a bit too grand for them. In reality it was not, accustomed as they were to staying in such hotels, but it was too extravagant for the occasion. As Carl and Emma knew, Freud had to earn every penny he spent providing for his large family. Besides, as Jung pointed out, the Regina was much closer – just round the corner from Berggasse.

This second visit to Vienna had undercurrents. Freud was still determined that Jung would be his crown prince, now replacing his usual *'Lieber Herr Kollege'* with *'Lieber Freund'*, a most informal address for those formal times. On other occasions it was even 'Dear Friend and Heir'. Jung progressed from *'Hochverehrter Herr Professor'* – most highly esteemed – to plain *Lieber Herr Professor*. But Jung still had his doubts. Ernest Jones, recalling Freud's failure to see this at the time, decided it was because Freud was so fixed on Jung becoming his 'heir' plus editor of their journal and president of the planned International Association, that he was blind to his doubts. This was only natural, Jones suggested, because:

> to begin with, Jung, with his commanding presence and soldierly bearing looked the part of a leader. With his psychiatric training and position, his excellent intellect and his evident devotion to work, he seemed far better qualified for the post than anyone else. Yet he had two serious

145

disqualifications for it. It was not a position that accorded with his own feelings, which were those of a rebel, a heretic, in short a 'son', rather than those of a leader, and this consideration soon became manifest in his failure of interest in pursuing his duties. Then his mentality had the serious flaw of lacking lucidity.

Jones – who would become an ardent follower of Freud, not averse to orchestrating oblique attacks on Jung – was at this early stage an astute observer. 'It was natural that Freud should make much of his new Swiss adherents, his first foreign ones and, incidentally, his first Gentile ones,' he wrote. 'After so many years of being cold-shouldered, ridiculed, and abused it would have needed an exceptional philosophical disposition not to have been elated when well-known University teachers from a famous Psychiatric Clinic abroad appeared on the scene in whole-hearted support of his work.' Freud was quite straight-forward about it: 'My selfish purpose, which I frankly confess, is to persuade you to continue and complete my work by apply-ing to psychoses what I have begun with neuroses. With your strong and independent character, with your Germanic blood which enables you to command the sympathies of the public more readily than I, you seem better fitted than anyone else I know to carry out this mission.' But Jung did not want to complete Freud's work. He wanted to complete his own.

The signs were already there six months earlier, in September 1908, when Freud came to visit Jung for four days at the Burghölzli. Emma was not there for most of it: she was preg-nant with their third child and spent the summer at Schaffhausen, not returning to Zürich until October.

Freud arrived at the Burghölzli in high good humour. He had been to Manchester to visit his half-brother Emmanuel, then on to London where he bought a fine new pipe, some excellent cigars, visited the National Gallery and the British Museum with its incomparable collection of antiquities. He sailed from Harwich via the Hook of Holland, stopping off in The Hague to see the Rembrandts. In Zürich he and Jung walked and talked for hours as usual, observed some of Jung's patients, and went climbing in the Alps – Mount Pilatus and the Rigi – each man clad in alpine boots and knickerbockers with their walking sticks and alpine hats. There was little sign of the illnesses which had been plaguing Jung that year, nor of the earlier tensions between Carl and Emma which were surely the cause. Freud said he was looking forward to visiting them in their new house; Jung said nothing could delight them more. It is likely that Emma was present for a day or two because, along with a thankyou letter, Freud sent her 'a surprise package of books', which suggests a conversation. 'Please tell your wife that one passage in her letter gave me particular pleasure,' he wrote to Jung after receiving Emma's letter. Later, during a short stay in Venice, he sent her a postcard. It marks the start of a relationship between the wise old Viennese professor and the young Frau Doktor Jung, who was just beginning to assert herself against an overpowering husband. And it sent a message to Jung: value her.

But for most of Professor Freud's visit Emma stayed with her mother at Ölberg. The weather that summer was exceptionally warm, with temperatures regularly reaching the mid-thirties. Luncheon was often served under the shade of one of the trees, afternoon tea on the veranda. The children played in the garden for hours, watched over by Emma and her mother knitting or doing their needlework. In the afternoons there was usually a walk through the meadow and up the hill behind the house into the forests beyond. Sometimes Marguerite joined them. She was also pregnant and living not far away in her own home – a house she and Ernst had commissioned in the latest modern style, in the sharpest contrast to Emma and Carl's traditional house at Kusnacht. Towards evening the hot weather could turn heavy at Ölberg, only broken by violent thunder storms when the maids ran about the house closing the windows and fastening all the shutters.

Before dressing for dinner, there was bedtime for the little girls, which Emma liked to do herself with the help of the children's maid. If Bertha was hosting one of her receptions, inviting all sorts of interesting people from the town and further afield – writers, astrologers, artists, often including the musicians from the town orchestra who performed a musical entertainment after the dinner – then Emma put on her jewels – pearls, rubies, diamonds – the ones she would show to her wide-eyed grandchildren years later. Emma always loved being in her mother's company. As the word association tests revealed, she could talk to her about anything, even some of the problems in her marriage. After all, it was Bertha who had encouraged her to marry Carl in the first place. Her mother, modern and broad-minded in outlook, knew the rumours about Carl's infatuations and flirtations and she knew what a complicated man he was, but she counselled fortitude. Her own marriage had not been easy, and Marguerite did not have an easy marriage either. Ernst Homberger had a reputation as a harsh, ambitious man who brought up his children with military discipline.

Carl only visited at the weekends, sometimes only on Sunday, as he was working his usual long hours. But he and Emma wrote to each other most days, a habit they maintained for the rest of their lives whenever they were apart. That summer he was preparing the first edition of the *Jahrbuch* and his own paper for it, 'The Significance of the Father in the Destiny of the Individual', as well as standing in for Bleuler during his summer vacation, attending court once a week to give psychiatric evidence, and receiving visits of doctors from foreign parts, including one from the president of the 'Lunacy Commission of New York State'. He also had to complete his military service, which ran to five weeks that year. His own holidays began on 21 August and he decided to spend the first part on his own 'fleeing into the inaccessible solitude of a little Alpine cabin on Mount Santis', as he put it to Freud. For just these few days the fine weather broke, the mornings so dense with fog there were no Alps to be seen and it rained for part of every day. But Carl had not come for the weather; he'd come for the climbing and the solitude. Emma was well enough acquainted with the 'other Carl' by now to know that solitude was indispensable to him. By the time he got to Schaffhausen there were only six days of his holiday left.

At the Burghölzli, one of the patients Freud and Jung observed was 'Babette S'. As Emma could have told Professor Freud, Babette S – old and demented as she was – was extremely important to Carl. He spent hours with her trying to discover meaning in her ramblings. She was the subject of his over-subscribed city hall lecture and she appeared in his *Psychology of Dementia Praecox*. 'She came out of the Old Town of Zürich, out of narrow, dirty streets where she had been born in poverty-stricken circumstances and had grown up in a mean environment,' he wrote. 'Her sister was a prostitute, her father a drunkard. At the age of thirty-nine she succumbed to a paranoid form of *dementia praecox*, with characteristic megalomania. When I saw her, she had been in the institution for twenty years.' Babette uttered and wailed the most extraordinary things. 'I am the Lorelei'; 'I am unjustly accused like Socrates'; 'I am the double polytechnic irreplaceable'; 'I am plum cake on a cornmeal bottom'; 'I am Germania and Helvetia of exclusively sweet butter'; 'Naples and I must supply the world with noodles'.

The reason for Jung's dedicated interest, as well as his evident compassion, was clear: 'My preoccupation with Babette and other such cases convinced me that much of what we had hitherto regarded as senseless was not as crazy as it seemed. More than once I have seen that even with such patients there remains in the background a personality which must be called normal. It stands looking on, so to speak. Occasionally, too, this personality – usually by way of voices or dreams – can make altogether sensible remarks and objections.' He knew from his own experience how thin the line between sanity and insanity could be. But all Freud had to say after the observation was: 'You know, Jung, what you have found out about this patient is certainly interesting. But how in the world were you able to bear spending hours and days with this phenomenally ugly female?' It was a double miss. Freud missed the reason why Jung spent all those hours with his schizophrenic patients; Jung completely missed his friend's Jewish humour.

There was another double miss on one of the evenings during the Jungs' second visit to Vienna. Emma had returned to the

Hotel Regina, alone as usual, whilst the two men sat talking late into the night in Freud's small study crammed with books and antiquities he had collected on his travels. On this occasion the conversation turned to the occult, precognition and parapsychology. The occult was still a live subject in Switzerland and Bleuler at the Burghölzli did not exclude it as a valid area of scientific investigation, even dispatching Jung to observe seances. However, Freud, Jewish, urbane and determinedly rational, rejected it out of hand. At this rejection, Jung felt a pressure in his diaphragm like red-hot iron, whereupon there was a loud 'report' from the bookcase, leaving both men shocked. Jung told Freud it was a 'catalytic exteriorisation phenomena'. 'That is sheer bosh,' replied Freud, whereupon an offended Jung predicted there would soon be another, which there was. Later Freud accounted for it in a rational way but at the time he was visibly shaken.

By the time Emma and Carl got back from Vienna the house at Kusnacht was almost ready, but not quite. They gathered up their children and the maid and went to stay at Schaffhausen again for the duration, Carl taking the opportunity to go off on a bicycle tour of northern Italy with an old friend. He had just done what he had been talking about for a long time: handed in his Burghölzli resignation to the Zürich authorities, which had been accepted, and now he felt free as air.

Things were not as they seemed, however. In fact Bleuler had dismissed Carl from his post the previous October, finally fed up with his arrogant behaviour and increasingly critical attitude, especially of Bleuler himself. Jung could spend hours with his patients, but still could not be bothered with the day-to-day administrative side of his job. Jung was shocked by his dismissal. That was not the way things were meant to be and it was not the way he meant to leave. He wanted to be affiliated to the Burghölzli so he could continue seeing his private patients and continue his lectureship at the university. Bleuler, ever generous, agreed to a compromise: Jung could stay in his post until the house at Kusnacht was ready, but then he had to go. They would put out a formal statement saying Herr Doktor Jung was leaving by mutual consent to pursue his private practice.

The intervening months were not a problem for Carl, but they were very difficult for Emma. She and Hedwig Bleuler had been friends since Emma first arrived at the Burghölzli in 1904. They had spent many hours together watching over the children and talking, and they had much in common: both married to men who had chosen to make psychiatry their lives, both mothers of growing families, both intelligent, keen to further their education. They had attended the Burghölzli meetings together and, as A. A. Brill described, were just as keen as the men to interpret dreams, analyse complexes, describe fantasies. The older woman had become something of a role model for Emma, introducing her to the idea that women could be ambitious, all the while supporting their husbands as society required. Emma helped Hedwig with the asylum's social events, and Hedwig's boys often ran upstairs to the Jungs' apartment to play with Agathli and Gretli. There is a charming out-of-focus photograph of Emma in the snow outside the front entrance of the Burghölzli with the children, perhaps Hedwig's, perhaps her own. Her skirts are long and dark and she is wearing a large hat, even though it is no more than an ordinary day. When the Jungs left the Burghölzli, the boy, Manfred, stood by the window of the Bleuler apartment with tears running down his face: he loved Frau Doktor Jung. But now Hedwig turned cold. She disapproved of Carl's treatment of her husband and she was not inclined to indulge Emma either. Gossip about Carl's flirtations did not help, nor the fact that the Jungs were so wealthy he did not have to work another day if he did not want to. Try as she might, there was nothing Emma could do. Her friendship with Hedwig was lost.

By the end of May 1909 Carl and Emma were finally able to move into the new house. 'This last miserable week should now be followed by festive days, for it really was a bad week, only my wife has kept her head above water,' Jung wrote to Freud from Kusnacht on 2 June. 'I am just beginning to guide my thoughts back into rectilinear channels, until now I haven't been able to concentrate on a thing. Although we began the move last Tuesday, only four rooms are really finished today. In the dining room, for instance, not even the floor is ready.' It did not stop him working though. 'Out here my practice is picking up again – something I hadn't expected.' By 12 June he was able to write: 'Today my children have moved into the new house. Everything is going well, including my practice, which makes me very happy.'

One of his new private patients was Joseph Medill McCormick, scion of the wealthy and powerful American newspaper proprietors of the *Chicago Tribune*. Medill was an alcoholic who had experienced a breakdown whilst travelling in Europe. Jung saw it as a providential act of fate that he landed on his doorstep at just the moment when he needed him most, setting up his private practice. But as so often with Jung, things were not what they seemed; in fact it was another reason Bleuler was fed up with him. Medill had ended up at the Burghölzli because it was an institution of international repute, not specifically for Herr Doktor Jung. But Carl bagged him before anyone else could get a look-in. Medill's problem, he quickly divined, was a powerfully dominating mother. When Medill returned to Chicago apparently 'cured', word went round about the remarkable Swiss doctor, and Medill joined the long line of wealthy Americans crossing the Atlantic to be treated by Jung.

In the intervening days, Emma and her newly engaged staff – a cook, a kitchen maid, two housemaids, the children's maid, one handy-man-cum-gardener – worked hard arranging all the furniture from the apartment at the Burghölzli, including the beautiful walnut desk, the tapestries and the paintings, and bringing some extra pieces from Ölberg as well. Everything again had to be moved by horse and cart, and the chaos was indescribable. The weather was unusually hot for that time of year, with the bluest skies and the Alps so clear you could almost touch them. That was at the beginning of the ten days' move. But then it turned to rain and thunderstorms. By the time the good weather came back the move was done and only the mess and muddle in the unfinished rooms remained.

On the ground floor the large tiled entrance hall was ready, the garderobe to the left for coats and hats, outdoor shoes and boots and a large collection of umbrellas and walking sticks, with a downstairs toilet next to it, modern and self-flushing, not like the majority of toilets in Switzerland at that time, which consisted of nothing more than a hole in a plank of wood and a deep shaft. A small salon led off to the left, a cosy room with dark wallpaper and a high tiled stove, bright blue, beyond which a heavily bolted door led out to the garden room, the one which had briefly seemed too expensive. The wooden stairs which led up to the next floor were wide and gracious befitting the tower which contained them. The kitchen to the right of the entrance was ready with a fine cooking range and a large pantry off, but the children's room next to it was not ready, nor, as Jung noted, was the pièce de résistance, the *Stube*, the long family room, with a brown marble fireplace flanked by two tapestry-covered chairs and the Frans Hals copy above, the one commissioned by Jung with a loan from his future mother-in-law when he was living in Paris. The alcove next to the fireplace was lit by two panes of stained glass, reproductions of medieval ones in Zürich's Landesmuseum: one commemorating Carl and Emma's marriage, the other Agathe and Gretli's births. One of Emma's large tapestries was hung at the other end, above a canapé with an antique dresser alongside. And her beautiful writing desk. In time Emma's piano came from Ölberg. Three long, deep-set windows overlooked the lake, glass cabinets in between with decorative landscape paintings above. Not so charming was the dark painting of John the Baptist's dripping head held by some unknown hand, which came from Jung's old parsonage home. A dining table and six chairs stood in the middle of the room on a Persian rug. From then on everything happened around that table: mealtimes, homework, drawing and model-making, knitting and stitching, and endless games of cards and mahjong.

Upstairs there was a spacious landing leading to the library, with Carl's *Cabinet* (study) and a small waiting room for the patients to the right, already finished so he could continue working undisturbed by all the surrounding chaos. The library is much the same today as it was then, Jung's books back in their original place on the shelves. One glance is enough to confirm Emma's hand: no dark wood, nothing medieval or eighteenth-century in the decor, but a spacious, light room overlooking the lake, the woodwork painted pale green with a white-and-green-tiled stove by the door. Like the family room it was parquet-floored. One wall was papered orange-brown, with a canapé and table below and a large painting of a young boy, naked, riding a white stallion bareback, above. The boy has his back to us, his arm raised in greeting to an unseen person out of frame, a poetic image probably painted by local artist Rudolf Koller. There is a communal vestibule between the library and Jung's *Cabinet*, with heavy double doors to shut out any noise and all connection to the outside world. Inside, the *Cabinet* has the feel of a chapel, small and dimly lit by three narrow, leaded windows each with a stained-glass panel representing the most Christian of iconographies: 'the Flagellation of Christ', 'the Crucifixion of Christ' and 'the Entombment'. No one could enter this sanctum without knocking, not even Emma.

The mood of the reception rooms might be traditional, but the rest of the house was thoroughly functional and modern, lit by electricity and heated by a central heating system fed by a huge wood- and coal-fired boiler in the cellar, installed by Gebrüder Sulzer of Winterthur. To the left of the upstairs land-ing beyond a glass door were three bedrooms, the one belonging to Emma and Carl having an en-suite bathroom. From the small corridor between the parents' and the children's bedrooms there was an exit to the roof of the veranda where the maids hung out the washing on washing days. One floor further up in the attic were the maids' and guest rooms.

Once the house was finished and everything in its place, Carl and Emma commissioned a local photographer to take pictures of the main rooms and the garden. You have to look closely at a small photo, also of 1909, taken of the terrace at the back of the house overlooking the lake, to discover the figures of two women. They are standing at the south corner of the terrace, one looking out, the other across, together but alone, posing for the camera. Emma and Trudi, Carl's sister. Trudi is wearing a long white dress, with her hair loosely rolled up. Emma wears her dark hair in a soft chignon at the nape of her neck, lovely in a quiet sort of way, as she was herself.

The terrace, 228 Seestrasse, 1909.

8

A Vile Scandal

Before they moved into the new house there were two events which took up a great deal of Emma's time and concern, the one happy, the other quite the opposite.

To start with happiness: Franz Karl Jung was born on 28 November 1908, their third child. Although Emma had spent the four months before the birth at Ölberg she was determined to have her baby at home, with Carl in attendance. It was six months before they could move into the house at Kusnacht, so back to the Burghölzli she came, together with the two little girls and the maid. And this time, to Emma's joy, it was a boy. 'Having dropped all my duties today because my wife is about to be confined, I at last have time to write to you,' Jung wrote to Freud on 27 November, to which clever Freud answered: 'Fate has made you a father again, and perhaps the star you spoke of on our long walk has risen for you. The transfer of one's hopes to one's children is certainly an excellent way of appeasing one's unresolved complexes, though for you it is too soon. Let me hear from you; until then I assume that the valiant mother is well; to her husband she must indeed be more precious than all her children, just as the method must be valued more highly than the results obtained by it.' Bertha was evidently not the only one who had heard rumours of Carl's flirtations.

'Heartiest thanks for your congratulatory telegram!' Jung responded. 'You can imagine our joy. The birth went off normally, mother and child are doing well. Too bad we aren't peasants anymore, otherwise I could say: now that I have a son, I can depart in peace.' And in a later letter: 'Everything is fine here. My wife is, of course, nursing the child herself, a pleasure for both of them.'

A month later he was writing about four-year-old Agathli's reaction to her baby brother's birth. When Jung asked her what she would say if the stork brought her a little brother, she replied: 'Then I shall kill it.' 'Franzli' had been born in the middle of the night. The next morning Jung carried Agathli to her mother's bedside. Emma naturally looked tired and pale, which alarmed the child, who said nothing and showed no joy. Later that morning she flung her arms round Emma: 'But, Mama, you don't have to die, do you?' A day later the two little girls went back to Schaffhausen to stay with their grandmother for several weeks. When they got back Agathli at first acted shy and suspicious of Emma, but once she had regained her confidence she became very fractious, asking endless questions: 'Shall I become a woman like you? Shall I then still talk to you?' 'Do you still love me too, not just Franzli?' 'Is that true? You're quite sure it's true? You're not lying? I just don't believe it.' Emma tried to distract her: 'Come, we'll go into the garden.' Agathli had become obsessed with news of an earthquake at Messina in Italy, '75,000 dead', demanding to be told about it again and again. Emma had to reassure her 'hourly' that there were no earthquakes in Zürich. Her father did the same.

Outside heavy snow fell. Emma was up and about again, finding books with pictures of earthquakes and volcanoes for Agathli, who pored over them, examining every detail. But she would not let it go. Finally Carl suggested that Emma tell Agathli how babies are made. Children grow in the mother like flowers on plants she explained, the days of sex education lying far in the future. It seemed to do the trick. 'Have you a plant in your tummy too?' she asked Carl and skipped off merrily when he said no. The next day she announced: 'My brother is also in Italy and has a house made of glass and cloth and it doesn't fall down.' When female guests came to the house she surprised them by asking whether they had a child or whether they had been to Messina. Then she invented a new game with her doll, sticking it between her legs under her skirt so only the head showed: 'Look, a baby is coming!' she cried, pulling it out slowly: 'And now it's all out!' One morning she insisted on coming into her parents' bedroom and created a scene when she was not allowed. Once Emma and Carl got up she hopped into the bed, lay flat on her stomach, and flailed about, kicking her legs like a horse: 'Is that what Papa does? That's what Papa does, isn't it?'

Meanwhile Gretli, aged three, was quite unperturbed by the birth and ridiculed the stork theory because the stork had not just brought Franzli, it had brought the children's nurse as well.

From all this one might assume that Carl was as happy to have a son as Emma. Not so. Six days after the birth he wrote a letter in sharp contrast to those he wrote to Freud, revealing a man in turmoil. 'I regret so much; I regret my weakness and curse the fate that is threatening me. I fear for my work, for my life's task, for all the lofty perspectives that are being revealed to me by this new *Weltanschauung* [world view] as it evolves,' he wrote on 4 December. 'You will laugh when I tell you that recently *earlier and earlier childhood memories have been surfacing*, from a time (*3rd to 4th year*) when I often hurt myself badly, and when, for example, I was once only just rescued from certain death by a maid. My mind is torn to its very depths, I who have to be a tower of strength for many weak people, am the weakest of all. Will you forgive me for being as I am? For offending you by being like this, and forgetting my duties as a doctor towards you? Will you understand that I am one of the weakest and most unstable of human beings?'

The recipient of this letter was Sabina Spielrein, she having made contact with Carl again in June of that year, and it became the prelude to the second event, the unhappy one, which eventually forced Emma to take the kind of action which would normally have horrified her. Ostensibly Fräulein Spielrein had written to Herr Doktor Jung to commiserate when she heard about the dramatic failure of his treatment of Otto Gross, the patient who had been a psychiatrist himself and had helped Jung with his paper about the 'Significance of the Father', suffering from a father complex himself, just like Jung. It did not take much to work out that Fräulein Spielrein, now completing her medical exams at the University of Zürich, was still obsessed with her former doctor.

Otto Gross had been passed to Jung by Freud in the hope that he could cure him of his opium and cocaine addiction before Freud took over further treatment, admitting to Jung that he himself had been dangerously addicted to cocaine as a young man, which, 'as I well know, produces a toxic paranoia'. In Freud's view Gross was the only other member of the psychoanalytic fraternity who could rival Jung in brilliance. But Jung soon discovered that Gross's problems extended beyond his addictions; he was suffering from a form of schizophrenia as well. He was anarchic and charismatic and before Jung knew it he was completely caught up in the case, spending hour upon hour analysing him. And when that did not work, he decided to reverse the roles, allowing the patient to analyse the doctor. This was dangerous territory for Jung – with his split personality and instability – and it made him vulnerable to all sorts of doubts and confusions.

Gross's schizophrenia stemmed from his childhood and a homosexual fixation on his father, which had led to obsessive masturbation from an early age. For Jung it was too close for comfort. 'In Gross I experienced all too many aspects of my own nature, so that he often seemed like my twin brother,' he later admitted to Freud, adding 'but for the *dementia praecox*' almost as an afterthought. Gross was serially unfaithful to his wife and as a self-proclaimed revolutionary, argued that political and sexual freedom went hand in hand, preaching polygamy as the only solution to marriage. Jung, who less than a year earlier had insisted to Freud that sexual repression was 'a very important and indispensable civilising factor', was soon sucked into this *Weltanschauung*. Heady with the thrill of their marathon conversations, and convinced of the apparent success of the treatment, Jung gradually reduced Gross's medication. It took a few weeks to realise that this was a disastrous mistake, by which time it was too late. One day in June, Jung had to admit to Freud that Gross had absconded over the Burghölzli's walls and disappeared, declaring himself 'cured'. Jung was left feeling guilty and confused, all sorts of demons reawakened. And, with

Emma away during the summer and then taken up with the birth of Franz, he was open to all sorts of temptations.

Which is when Sabina Spielrein wrote her letter to Carl, making contact again.

'My dear Fräulein Spielrein,' he replied. 'You have managed well and truly to grasp my unconscious with your sharp letter,' referring bitterly to the Gross debacle. 'Such a thing can only happen to me.' He suggested they meet at the paddle steamer landing stage on the Bahnhofstrasse at 11 a.m. the following Tuesday: 'So we can be alone and able to speak undisturbed, we'll take a boat on the lake. In the sunshine, and on the open water, it will be easier to find a clear direction out of this turmoil of feelings. With affectionate greetings from your friend.'

What was Carl thinking? He knew better than anyone how vulnerable Fräulein Spielrein was and how dangerous it was to take up contact again, but the 'other' Carl needed someone to talk to, and Emma was not the one. The birth of Franz, a son and heir, only made matters worse. Now 'The Significance of the Father in the Destiny of the Individual', based on his experiences as a son, referred to him as a father as well. Carl, 'other' Carl, felt no joy, only regret and fear of instability, his mind 'torn to its very depths'.

Seventy-five years later a suitcase belonging to Sabina Spielrein was found in an attic in Geneva. It contained twenty-one yellowing double pages of small black handwriting – a journal, undated, sometimes coherent, often rambling and opaque, the outpourings of a desperate young woman in a foreign land still traumatised by her father's sexual abuse and still obsessed with her one-time doctor, Jung. In addition, there are letters, undated and not always complete, which Spielrein later sent to Freud describing and sometimes quoting from letters Jung wrote to her. And some dated letters from Jung and Freud. There is also a diary from 1909 to 1912. The problem this material presents is that, apart from the Jung and Freud letters, all the evidence comes from Spielrein herself, with no one to corroborate it. The journal is so rambling and oblique that it is impossible to work out what is fact, what is wish-fulfilment or fantasy, leaving others to interpret it as they may.

Jung's letter to Fräulein Spielrein six days after Franz's birth was confused and desperate. Apart from memories of childhood which kept coming back to him unbidden, he was paranoid, fearing Fräulein Spielrein would take revenge on him once she discovered how unstable he was. He was looking for someone 'as yet unrealised' who would love him with no strings attached, social or other. And no punishment. 'It is my misfortune that I cannot live without the joy of love, of tempestuous, ever-changing love. This daemon stands as an unholy contradiction to my compassion and my sensitivity. When love for a woman awakens in me, the first thing I feel is regret, pity for the poor woman who dreams of eternal faithfulness and other impossibilities, and is destined for a painful awakening out of all these dreams.' As the months passed Spielrein became even more obsessed with Jung. She knew he would never leave Emma, but she held fast to the idea that he loved her. Then she had decided she wanted a child by him. That brought Carl sharply back to reality. The last thing he wanted was a child by Spielrein. He wanted an infatuation. And someone to talk to about his torments.

At some point in early 1909 Emma realised Carl was seeing Fräulein Spielrein again and it confronted her with the greatest dilemma of her life so far. Married for six years, the mother of three children and in love with a husband who could not stop flirting with other women, she was still too young and inexperienced to see the situation for what it was, still dreaming of 'eternal faithfulness and other impossibilities'. She became so desperate she took an action utterly alien, even repugnant, to her: she wrote an anonymous letter to Fräulein Spielrein's mother, warning her to 'rescue' her daughter before Doktor Jung 'ruined' her. We know about it because of a letter Fräulein Spielrein wrote to Freud in June 1909: 'Four and a half years ago Dr Jung was my doctor, then he became my friend and finally my "poet", i.e. my beloved,' she wrote. 'Eventually he came to me and things went as they usually do with "poetry". He preached polygamy; his wife was supposed to have no objection, etc. etc. Now my mother receives an anonymous letter that minces no words, saying she should rescue her daughter, since otherwise she would be ruined by Dr Jung. The letter could not have been written by one of my friends, since I kept absolutely mum, and always lived far away from all the other students. There is reason to suspect his wife (?)' she wrote.

In fact rumours of Jung's relationship with a young lady had already reached Freud. He sent Jung a telegram which put him 'in a fluster'. Jung replied that he had not written for such a long time because he had been 'under terrific strain day and night', blaming a mass of correspondence every evening, plus invitations, concerts, three lectures, and all the while the new house. 'The last and worst straw' he wrote:

is that a complex is playing Old Harry with me: a woman patient, whom years ago I pulled out of a very tricky neurosis with unstinting effort, has violated my confidence and friendship in the most mortifying way imaginable. She has kicked up a vile scandal solely because I denied myself the pleasure of giving her a child. I have always acted the gentleman towards her, but before the bar of my rather too sensitive conscience I nevertheless don't feel clean and that is what hurts the most because my intentions were always honourable. But you know how it is – the devil can use even the best of things for the fabrication of filth.

This is surely Jung at his worst, spouting excuses, feeling sorry for himself, indignant even. Reading between the lines it sounds as though Jung went much further with the relationship than was 'proper', but not so far as to become what Fräulein Spielrein enigmatically called her 'poet'. A case of the spirit if not the letter of the law. 'I have always told your daughter a sexual relationship was out of the question,' Jung wrote to Fräulein Spielrein's mother after she had received the anonymous letter, leaving the question of 'poetry' up in the air.

There had been a row with Emma. 'Meanwhile I have learnt an unspeakable amount of marital wisdom,' his letter to Freud continued, 'for until now I had a totally inadequate idea of my polygamous components despite all self-analysis. Now I know how and where the devil can be laid by the heels. These painful yet extremely salutary insights have churned me up hellishly inside, but for that very reason, I hope, have secured me moral qualities which will be of the greatest advantage to me in later life. The relationship with my wife has gained enormously in assurance and depth.' Characteristically he failed to praise Emma for her part in it, though he knew well enough how much courage it must have taken for her to confront him, rescuing him from himself. Instead he talked on about his own part in it: he did not yet possess Freud's assurance and composure, he explained: 'Countless things that are commonplace for you are still brand new experiences for me, which I have to relive in myself until they tear me to pieces. This urge for identification (at the age of eleven I went through a so-called traumatic neurosis) has abated considerably of late, though it still bothers me from time to time.' He turned quickly to other matters, the journal, Ernest Jones who had lately become sexually involved with Otto Gross's wife, and finally: 'My small son is flourishing, my wife is in good shape.'

Jung need not have worried about Freud, who wrote back happily about an invitation he had received from America to give a series of lectures at Clark University in Worcester, Massachusetts, and telling Jung that he had already heard about the young woman who was going about introducing herself as the mistress, which he did not believe and put down to the neurosis of the woman in question. 'To be slandered and scorched by the love with which we operate – such are the perils of our trade, which we are certainly not going to abandon on their account.' There was no chance that Freud, still determined to have Jung as his crown prince, would side with the unknown woman, an ex-patient to boot. He was more bothered by Jung's talk of the devil and filth. 'You definitely lapse into the theological style in relating this experience,' he wrote, obliquely alluding to Jung's father, the pastor of the Protestant Reformed Church. Jung, usually so dilatory, replied immediately: 'I must answer you at once. Your kind words have relieved and comforted me.' He reassured Freud about his 'theological' style, not noticing Freud's warning: that Jung was still labouring under a father complex, an opinion Freud never altered, believing it lay at the heart of Jung's neuroses. 'Now and then, I admit, the devil does strike a chill into my – on the whole – blameless heart,' Jung wrote shiftily. But he denied the rumour outright: 'I've never really had a mistress and am the most innocent of spouses.' He simply could not imagine who the woman might be. And that, he hoped, was that.

By June, the Jungs had settled into their house at Kusnacht – 228 Seestrasse – and a new era began. It was part of the agreement Emma reached with Carl after the showdown: new house, new life. On one level it worked beautifully: this was a place they had created together and they both loved. Emma knew this was one thing Carl would never want to give up: this home, this stability.

'Hurrah for your new house!' wrote Freud on 3 June 1909. But in the same letter he enclosed another, from Fräulein Spielrein. 'Weird! What is she? A busybody, a chatterbox, or a paranoiac? If you know anything about the writer or have some opinion in the matter, would you kindly send me a short wire, but otherwise you must *not* go to any trouble.' Jung shot off a telegram, immediately followed by a letter. 'Spielrein was the person I wrote you about,' he finally admitted, explaining that he had prolonged the relationship because he felt indebted to her for being the most important of his early 'test-cases' and because he was worried that she was not fully cured and might relapse. 'She was, of course, systematically planning my seduction, which I considered inopportune. Now she is seeking revenge. Lately she has been spreading a rumour that I shall soon get a divorce from my wife and marry a certain girl student, which has thrown not a few of my colleagues into a flutter. What she is now planning is unknown to me. Nothing good, I suspect.' Hers was a case of 'fight-the-father', just like Gross, he explained, which he had been trying to cure with 'untold tons of patience, even abusing our friendship for that purpose', and the whole thing had set off a complex in him like his 'compulsive infatuation' with the Jewish woman in Abbazia after their fist visit to Vienna.

Freud wrote to Fräulein Spielrein making it clear whose side he was on. To Jung he wrote: 'Such experiences, though painful, are necessary and hard to avoid. Without them we cannot really know life and what we are dealing with. I myself have never been taken in quite so badly, but I have come close to it a number of times and had *a narrow escape*.' The fact that he was older than Jung helped and no lasting harm had been done. 'They help us to develop the thick skin we need and to dominate "counter-transference" which is after all a permanent problem for us; they teach us to displace our own affects to best advantage.' They are a *'blessing in disguise'*. Both 'narrow escape' and 'blessing in disguise' are written in English. The fact is, for both Freud and Jung nothing mattered so much as their pioneering work and assuring the place of psychoanalysis in the future. And if it required a thick skin, well, so be it.

But Fräulein Spielrein was not quite done. She was too intelligent and resourceful, in spite of her emotional difficulties, to let the two men gang up against her in that way. On 19 June she turned up at the Jung house in Kusnacht insisting on talking things through. Emma had little choice but to let her in, and Carl had little choice but to come to terms. She insisted he write a letter to Freud. 'Although not succumbing to helpless remorse, I nevertheless deplore the sins I have committed, for I am largely to blame for the high-flying hopes of my former patient,' he duly wrote, describing Fräulein Spielrein as having freed herself from the transference 'in the best and nicest way' and admitting he had written letters to her mother Frau Spielrein, which he called a 'piece of knavery', all of which he 'confessed' to Freud 'as my father'. Freud cannot have relished the role of Jung's 'Father Confessor' and what it told him about Carl's continuing complexes.

But he did as Jung requested – 'a great favour' – he wrote to Fräulein Spielrein, confirming that Jung had written to him about the whole thing 'in perfect honesty' (this in English again). As for Jung, after asking Freud to 'pardon' him once again, he was breezily off on another matter: he too had now been invited to America to speak at Clark University, and he was happy to confirm he had managed to book a cabin on the same transatlantic liner as Freud and his colleague the Hungarian psychoanalyst Sandor Ferenczi, but 'a very expensive cabin, however', there being nothing else available at this late stage.

Life at 228 Seestrasse fell into place again. The Jungs' summers assumed a pattern which would last through the years: home-based, with Carl working all hours and only taking odd breaks for a bicycle tour or an Alpine hike with a friend. And sailing, now that he had his longed-for boat happily moored in his boat-house, fulfilling all his boyhood dreams and visions. Emma meanwhile ran the house, supervising the maids, the children's nurse, the cook and the gardener. She saw Carl's mother and sister often, since they lived nearby, and took the children off to Schaffhausen whenever possible.

On 13 July 1909 Jung wrote to Freud listing everything he had to do before leaving on the trip to America: the *Jahrbuch*, six private patients, the lectures for America, and writing up two reports for the courts – all of which cannot have left much time for Emma and the family. Freud replied from Ammerwald in the Austrian Alps, where the Freud family liked to spend their summers, and making a point of signing off with kind regards to Jung's 'charming' wife.

By 18 August, Carl was bound for Bremen to meet Freud and Ferenczi and embark on the liner the *George Washington*, briefly stopping off en route in Basel to visit old friends and family. Emma was probably relieved to see him go after the dramas with Fräulein Spielrein. But Carl's long letters kept her fully up to date throughout his seven-week absence. Keen psychoanalysts as they were, the men took the opportunity of analysing each other's dreams during the nine-day sea voyage, whiling away the time in the luxury of Jung's first-class cabin, or seated in a corner of the fine library or one of the plush salons. When Jung asked Freud for more detail from his private life to help with the interpretation of one of his more intractable dreams, Freud refused, saying he could not 'risk his authority!' That did it for black-and-white Carl: 'Freud was placing personal authority over truth.' Later he considered this refusal as a turning point, foreshadowing the break-up of their relationship. 'Father was very disappointed in his own father; and after that dream on the ship, he became very critical of everything that Freud said,' was Jung's son Franz's opinion years later, with the benefit of hindsight. 'He had a negative-father-complex, and he brought it to his relationship with Freud.'

On the voyage Jung recounted one of his 'house' dreams, the kind he had had ever since he was a child. It started on the top floor in the rococo style, then descended, the floors getting more and more ancient until he arrived at a heavy door to a vaulted cellar with a stone floor, where he discovered an old ring which he pulled to reveal some narrow stone steps leading down even further to a low cave cut into the rocks, with thick dust, scattered bones and broken pottery on the ground, the remains, he presumed, of a primitive culture. And two skulls. Jung later described it as one of his most important dreams, presaging his theory about the 'collective unconscious', but Freud focused exclusively on the two skulls, suspecting Jung of death-wishes against him – the father – and wanting to replace him. 'I knew perfectly well, of course, what he was driving at: that secret death-wishes were concealed in the dream,' wrote Jung later, admitting that he decided to go for the easy solution: he lied. He said they were the skulls of his wife and sister-in-law. 'After all, I had to name someone whose death was worth wishing! I was newly married at the time and knew perfectly well there was nothing within myself which pointed to such wishes . . . And so I told him a lie. I was quite aware that my conduct was not above reproach, but "*à la guerre, comme à la guerre!*"'

'*Liebste Frau!*', Dearest wife, Carl wrote to Emma from New York on 31 August 1909. 'When did I last write? I think it was yesterday. Time here is so frightfully filled up. Yesterday Freud and I spent several hours walking in Central Park and talked at length about the sociological problems of psychoanalysis. He is as clever as ever and extremely touchy; he does not like other sorts of ideas to come up, and, I might add, he is usually right.' Their relationship had evidently weathered the storms of the two dreams. After the walk they were invited to Dr Abraham Brill's home for supper. 'The meal was remarkable for the unbelievable, wildly imaginative dishes,' wrote Carl, wide-eyed:

> Picture a salad made of apples, head lettuce, celery root, nuts etc. etc. But otherwise the meal was good. Afterwards, between 10 and 12 pm, we drove down to Chinatown, the most dangerous part of New York, accompanied by three sturdy rascals. The Chinese all wear dark blue clothing and have their hair in long braids. We went into a Chinese temple, located in a frightful den called a joss house. Around every corner a murder might be taking place. Then we went into a Chinese teahouse, where we had really excellent tea, and along with it they served us rice and an incredible dish with chopped meat, apparently smothered in earth worms and onions. It looked ghastly. But the worms turned out to be Chinese potato, whereupon I tasted some, and it was not at all bad.

There were 'hoodlums' hanging about – 9,000 'Chinamen' to only twenty-eight women – and scores of white prostitutes, who had just been cleared out by the police. Next they went to 'a real Apache music hall . . . a rather gloomy place' where the audience threw money at the feet of the singer to show their appreciation. 'Everything most odd and terribly discomfiting, but interesting,' he confided to Emma. 'I should mention that Dr Brill's wife was along for the whole expedition, like the good American she is. We finally got to bed at midnight.'

The next morning, with his usual boundless energy, Jung was up at seven and off to visit Ward Island, the home of the New

York State Psychiatric Institute, then on to the Metropolitan Museum looking at the Egyptian, Cypriot and Cretan antiquities collections, as well as the 'wonderful' tapestries of the Pierpont Morgan Collection. He went alone because Freud and Ferenczi were probably still recovering from the journey and the exertions of the night before, and as the day went on he started to feel lonely and adrift in this strange land, telling Emma: 'Today there's some homesickness floating on the surface, sometimes no slight amount. I long for you and keep thinking whether you would like it here.' By the evening he had cheered up again. They were going out to Coney Island, the largest marine amusement park on the Atlantic coast, and he was also going to do a bit more work on his Worcester lectures, though it was impossible to concentrate in such a place and in such heat. Freud found the same, he said. After Worcester he planned to visit Niagara Falls, and perhaps travel some way into Canada but he thought he would skip Chicago for lack of time, 'because I want to be back on 21st September at the very latest. This can not go on too long. Give my best to everyone, and many kisses from your Carl.'

His host at Clark University was its president, the distinguished scholar G. Stanley Hall. Once settled in Professor Hall's house, Jung wrote to Emma describing their journey from New York. All three visitors had been suffering from diarrhoea and stomach pains, no doubt those Chinese 'worms'. They had taken the 'elevated' from 42nd Street to the piers, joined by Ernest Jones who had just arrived from Europe, and boarded 'a fantastically huge structure of a steamer that had some five white decks', sailing from the West River round the point of Manhattan with its 'tremendous skyscrapers' up the East River under Brooklyn and Manhattan Bridges and through the Sound behind Long Island. They arrived in Fall River City the next morning and took the train to Boston and thence to Worcester. The countryside was 'utterly charming, low hills, a great deal of forest, swamp, small lakes, innumerable huge erratic rocks, tiny villages with wooden houses, painted red, green or grey, with windows framed in white (Holland!)'. The happy reference was to one of their annual spring holidays together.

Dinner at the Halls' on that first evening was full of interest. Professor Hall himself was a 'refined', 'distinguished' old gentleman, his wife 'plump, jolly, good-natured, and extremely ugly'. You can almost hear Emma laughing as she read it. There were books everywhere and many boxes of cigars, which surely lightened Freud's heart. Freud suffered from a travel phobia, fretting about this and that – even wetting himself in agitation on one occasion – so arriving at the Halls' comfortable New England home must have come as some relief. 'Two pitch-black Negroes in dinner jackets, the extreme of grotesque solemnity, perform as servants. Carpets everywhere, all the doors open, even the bathroom door and the front door; people going in and out all over the place; all the windows extend down to the floor,' wrote Carl, knowing how interested Emma was in house details, especially now they had their own.

Two days later he wrote again. Freud had begun his lectures to great applause. 'We are gaining ground here, and our following is growing slowly but surely.' Solid Swiss Carl was surprised to find the American ladies were cultivated, well informed and free-thinking. He was happy to admit to Emma that he had been surrounded by five of them at a garden party. 'I was even able to make jokes in English – though what English!' One can imagine it: this large, brash, charismatic man, ruddy with alpine health, roaring with laughter at his own jokes and his awful English, with Freud nearby, small, dapperly dressed, just beginning to find his form in the new world, surrounded by other admirers, seducing them with his quieter, subtle Jewish–Viennese humour. They were causing quite a stir, were older Freud and young Jung, both of them interviewed by the *Boston Evening Transcript*. Things were going well. 'In fact we are the men of the hour here,' Carl declared. 'It is very good to be able to spread oneself in this way once in a while. I can feel my libido is gulping it in with vast enjoyment . . .'

Emma certainly knew what that meant. But out of sight was out of mind. Before he left Zürich, Carl had been in a parlous state: depressed, ill, falling into rages which he seemed unable to control. Along with everything else, he had been bypassed for a professorship at the University of Zürich for the second time. He blamed Bleuler, apparently not realising the faculty had its own reasons: Herr Doktor Jung's reputation with women went before him. But now he wrote: 'I have, thank God, completely regained my capacity for enjoyment, so that I can look forward to everything with zest. Now I am going to take everything that comes along by storm, and then I shall settle down again, satiated.'

The main thing both Jung and Freud meant to take by storm was America itself. In Carl's case, he had already treated some American patients in Zürich, and a handful of early practitioners of psychoanalysis in the United States, such as Abraham Brill, had made their way to the Burghölzli to observe him at his work and work there themselves for a time. Now he meant to build on his growing reputation, lecturing, giving interviews, writing papers, meeting distinguished Americans, and acquiring influential patients. One of these was Joseph Medill McCormick, whom Jung had first treated in Zürich earlier that year. 'Fate, which evidently loves crazy games, has just at this time deposited on my doorstep a well-known American (friend of Roosevelt and Taft, proprietor of several big newspapers etc.),' as he had written to Freud on 7 March 1909, right in the middle of the Spielrein drama.

Medill, as he was mostly known, was the grandson of Joseph Medill, publisher of the *Chicago Tribune*, the great-nephew of Cyrus Hall McCormick, founder of the International Harvester Company, and cousin of Harold Fowler McCormick, married to Edith Rockefeller. When Medill married Ruth Hanna, daughter of Mark Hanna, one of the Republican Party's big hitters, President Theodore Roosevelt himself travelled to Cleveland for the wedding, bearing a gold coffee set as a present. Medill was the manager and assistant editor-in-chief at the *Tribune*, but known mainly as a dedicated party-goer and an alcoholic. When he came to see Jung in Zürich it was the second of Medill's leaves of absence from the *Tribune*. Jung's treatment worked for a time, mainly because he insisted Medill quit his job, thereby escaping the clutches of his 'daemon' mother, who was a powerful figure on the *Tribune* board and the dominating force in his life. On this trip to America Jung saw Medill once, in New York. 'Jung was pleased with me apparently,' Medill wrote to his wife Ruth. 'He rejoiced that the savage in me looked out of my eye . . . and warned me against being too good . . . He said he quite permitted flirtations and had had one himself in this country.'

Medill also told Ruth that 'the doctor' had had the same 'conflicts' with his wife as himself: apparently Emma Jung used

to think 'she could not understand his science and paid no attention to it. Now she is his partner in his work.'

1909 is the year Emma began to move from being little more
than Carl's occasional assistant to being more of a partner in his
work, not that they shared the work equally, but in the sense
that Emma now knew enough about psychoanalysis to offer
suggestions and criticisms, and enough to pursue her own interest in the field. The day Fräulein Spielrein turned up at their
home had been a watershed. Emma finally found the courage to
put her foot down: she would no longer be the mere wife and
mother. They would work together again, as they had in the
early days of their marriage. Other husbands might have
demurred, but not Carl. He was always encouraging Emma to
do more for herself, develop her own work and her own life.

Carl's letters to Emma from America show the closeness of
their relationship and how far their partnership had come. He
wanted to tell her everything. '*Mein liebster Schatz*', my darling
treasure, he wrote from a camp in the Adirondacks, where he
had been invited by James Jackson Putnam, one of the guests at
Clark University, on 16 September at 8.30 in the morning. 'You
would be absolutely amazed if you could see where I have ended
up this time in this land of truly boundless opportunities':

> I am sitting in a large one-room wooden cabin looking into a
> massive fireplace of rough brick with mighty logs on the
> hearth. The walls are crowded with china, books, and the like.
> Around the cabin runs a covered porch, and when you step
> out the first thing that meets your eye is a sea of trees – beech,
> fir, pine, cedar, everything slightly eerie in the gentle rustling
> rain. Through the trees you can glimpse a mountainous land-

scape, all of it forested. The cabin stands on a slope, and somewhat further down you can see about ten other wooden cabins. Over here the women live, and over there the men; that is the kitchen, there you see the dining hall, and cows and horses are grazing among the buildings. I must explain that two Putnam families live here, and a Bowdrich family, complete with servants.

He describes primeval forests, glacial boulders, moss and fern, blackberry and raspberry bushes, wild, 'and a curious cross between the two'. Climbing to the top of a 3,500-foot bluff, he describes looking out over:

a wild glacial landscape of fields and lakes covered since the time of the glaciers with virgin forest . . . The area still has bears, wolves, deer, elk, porcupines. Snakes also abound. Yesterday when we arrived a two-footer was waiting to welcome us. My last letter was written to you in the railway station at Lake Placid, at the end of the line. From there we continued on to here, travelling for more than five hours in a curious two-horse conveyance over deeply rutted roads. All the Gerstacker memories [a German writer of American West stories] of my boyhood came rushing back. In what seemed like a completely desolate area we saw metal boxes nailed to trees so the mailman could drop off the letters for the farmers. Then came the little wooden shack by the road which housed the 'store', carrying every conceivable line of merchandise, then the 'hotel', where for lunch we were served 'brown bread' and 'corn on the cob' with salted butter and crisp bacon. A fire was lit. The Putnams have a harmonium; we sang German folk songs to it! They are terribly nice people. The hospitality is downright Indian. Except for train tickets I hardly need money. We really must come back here together some time; it is just too good to be missed . . .

A couple of days later he wrote again:

Two more days before departure! Everything is taking place in a whirl. Yesterday I stood upon a bare rocky peak nearly 5,600 feet high, in the midst of tremendous virgin forests, looking far out into the blue infinities of America and shivering to the bone in the icy wind, and today I am in the midst of the metropolitan bustle of Albany, the capital of the State of New York! The hundred thousand enormously deep impressions I am taking back with me from this wonderland cannot be described with the pen. Everything is too big, too immeasurable. Something that has been gradually dawning on me in the past few days is the recognition that here an ideal potentiality of life has become reality. Men are as well off here as the culture permits; women badly off. We have seen things here that inspire enthusiastic admiration, and things that make one ponder social evolution deeply. As far as technological culture is concerned, we lag miles behind America. But all that is frightfully costly and already carries the germ of the end in itself. I must tell you a great, great deal. I shall never forget the experiences of this journey. Now we are tired of America. Tomorrow morning we are off to New York, and on September 21 we sail!

9

Emma Moves Ahead

Carl arrived back from America in the highest spirits. There had been a storm lasting a day and half the night during their sea passage from New York to Bremen and he had spent much of it in a protected spot under the bridge of the *Kaiser Wilhelm der Grosse*, admiring the mountainous waves rolling over the deck until he got soaked through. When he went below decks for a cup of tea he immediately felt sick and quickly retired to his cabin, where he found all the objects had come to life, as he described in a letter to Emma the next day:

The sofa cushion crawled about on the floor in the semidarkness; a recumbent shoe sat up, looked around in astonishment, and then shuffled off quietly under the sofa; a standing shoe turned wearily on its side and followed its mate. Now the scene changed. I realised that the shoes had gone under the sofa to fetch my bag and briefcase. The whole company paraded over to join the big trunk under the bed. One sleeve of my shirt waved longingly after them, and from inside the chests and drawers came rumbles and rattles. Suddenly there was a terrible crash under my floor, a rattling, clattering, and tinkling. One of the kitchens is underneath me. There, at one blow, five hundred plates had been awakened from their deathlike torpor and with a single bold leap had put a sudden end to their dreary existence as slaves. In all the cabins round about, unspeakable groans betrayed the secrets of the menu. I slept like a top . . .

Emma, waiting Penelope-like at home, did not share his high spirits. Whether it was the usual discrepancy between the excited traveller and the dull stay-at-home, or the various warning signs which had crept in between the lines of Carl's letters, she was feeling low, made all the worse by Carl bounding in with his ebullient talk and loud laughter, turning everything upside down. He wanted to tell her everything – not least that he had put away the Burghölzli teetotalism for ever, having discovered the joys of champagne. 'As far as abstinence goes, I've arrived on very shaky ground indeed,' he had admitted in a letter, speaking of the drink but implying much else besides. 'I confess myself an honest sinner and only hope that I can endure the sight of a glass of wine without emotion – an undrunk glass of course. That is always so: only the forbidden attracts. I think I must not forbid myself too much . . .'

Faced with Emma's dissatisfied frame of mind, Carl decided the best solution was to analyse her. Or perhaps it was Emma who insisted. 'I feel in top form and have become much more reasonable than you might suppose,' Jung wrote on 1 October 1909 to Freud. 'My wife is bearing up splendidly under the psychoanalysis and everything is going *à merveille*.' He ended his perky letter by telling Freud about the man who invented muesli: 'In Zürich a Dr Bircher (please note the name!) has set up as a psychoanalyst. Formerly he believed in uric acid and apple sauce and porridge. Naturally he hasn't a clue.' Two weeks later Jung was still in the best of spirits. 'In my family all is well, thanks to lots of dream analysis and humour. The devil seems to be beaten at his own game.' After Christmas he and Emma went away for six days to Unterwasser, a secluded Alpine village in the canton of St Gallen. It was perfect: thick snow and bright winter sunshine.

So far so good. But Emma and Carl soon found it was not easy for a husband to analyse the dreams of his own wife, so in due course a colleague, Dr Leonard Seif, when he arrived from Munich to work in Zürich, was called in to take over. Evidently Emma was determined to tackle the dilemmas in her life and marriage. She had a fairly objective view of her husband's failings, Seif later told Ernest Jones, a man with plenty of marital failings of his own. Emma's problem was that talking about it and analysing her dreams did not make Carl's failings go away. Soon enough the same old devil raised its ugly head. As Carl explained to Freud on 30 January 1910:

During the time I didn't write to you [seventeen days] I was plagued by complexes, and I detest wailing letters. This time it was not I who was duped by the devil but my wife, who lent an ear to the evil spirit and staged a number of jealous scenes, groundlessly. At first my objectivity got out of joint (rule 1 of psychoanalysis: principles of Freudian psychology apply to everyone except the analyser) but afterwards snapped back again, whereupon my wife also straightened herself out brilliantly. Analysis of one's spouse is one of the more difficult things unless mutual freedom is assured. The prerequisite of a good marriage, it seems to me, is the licence to be unfaithful.

I in my turn have learnt a great deal. The main point always comes last: my wife is pregnant again, by design and after mature reflection.

Jung appears to have come out of the quarrels with Emma unbowed, holding fast to his new conviction that he should have a 'licence to be unfaithful'. Freud, answering the letter, made no comment about Jung's prerequisite of a good marriage, contenting himself with: 'I should have thought it quite impossible to analyse one's own wife.'

There was no lack of women to trigger Emma's 'number of jealous scenes': Fräulein Aptekmann, Maria Moltzer, Martha Boeddinghaus, all independent women connected with the Burghölzli, all vying for Carl's attention if not his love. And then there were the wealthy ladies of Zürich's *Pelzmäntel* brigade, who still thronged to his lectures. And Fräulein Spielrein. Emma probably thought the worst was behind her, but it was still worrying when, throughout the summer, Sabina Spielrein appeared at the house, ostensibly to ask for Carl's help with her dissertation, in fact trying to rekindle the flames of their relationship. How must it have felt for Emma to watch Fräulein Spielrein go up the stairs to Carl's *Cabinet* and not emerge for the next hour? Who knew what was happening behind those closed doors.

But Emma knew Carl and his 'infatuations'. As did Fräulein Spielrein. 'My friend said we would always have to be careful not to fall in love again; we would always be dangerous to each other,' Spielrein wrote ecstatically in her journal on a good day when the Herr Doktor 'pressed my hands to his heart several times and said this should mark the beginning of a new era'. But this was quickly followed by despair. 'My heart contracts painfully, for the main thing is missing, and this main thing is love. Oh, again this "What to do?" I hardly believe that I could love anyone the way I love my friend. I fear my life is ruined.' And a few days later: 'To be one among the many who languish for him, and in return receive his kind gaze, a few friendly words . . . fulfil his every wish . . . his vanity. He assumes a frigid, official tone, and who suffers from that? Not he, of course.' Out walking one day she bumped into Fräulein Aptekmann, another of Herr Doktor Jung's students, also besotted with him. 'She loves him and believes that he loves her,' wrote Spielrein that evening in her journal. 'She was so happy about it – her eyes gleamed, her cheeks glowed.' Emma could only watch as all these young women competed for Carl's attention and love, interpreting his flirtations as something more.

Using 'poetry' again to describe in typically enigmatic terms the 'more' in their relationship, Fräulein Spielrein wrote: 'Yes, the stronger poetry probably occurred a week ago. He said then that he loves me because of the remarkable parallelism in our thoughts . . . he told me he loves me more for my magnificent proud character, but he also told me that he would never marry me because he harbours within himself a great philistine who craves narrow limits and the typical Swiss style.' One day, Carl showed her sections of the secret diary he had kept until he married Emma, after which he had put it away in a drawer:

When he gave me his diary to read, he said in a very soft, hoarse voice, 'Only my wife has read this . . . and you.' And yet his wife, who, as his diary makes clear, hesitated for a long time before marrying him, because in spite of her love she

gave thought to her own comfort and did not want any wild-
eyed 'slave to an ideology'; his wife is protected by the law,
respected by all, and I, who wanted to give him everything I
possessed, without the slightest regard for myself, I am called
immoral in the language of society – lover, maybe maîtresse!
He can appear anywhere in public with his wife, and I have to
skulk in dark corners . . . True, he wanted to introduce me in
his house, make me his wife's friend, but understandably his
wife wanted no part of this business.

When Fräulein Spielrein turned up at the house for her appointment on 20 September 1910 she was turned away by a maid
without an explanation. Upstairs, in the bedroom overlooking
the garden, Emma had that morning given birth to a baby girl,
Marianne, their fourth child.

For Carl it was a repeat of his experience after the birth of
Franz: happiness closely followed by distress. Split Carl. When
she came for her postponed appointment two days later Fräulein
Spielrein found him in the kind of mood which gave her hope:
'Yes, my dear good friend, I love you and you love me. The
thing I was longing for only recently has been fulfilled: he
revealed his love almost too clearly . . . We were supposed to sit
down and work . . . So I am not one among the many . . . He
resisted, he did not want to love me. Now he must, because our
souls are deeply akin.' How could she know it was not about
her but all about Jung's own complex psychology? As he put it,
remembering how he had carried his own once powerful father,
now dying, from room to room: 'To be fruitful means, indeed,
to destroy one's self, because with the rise of the succeeding
generation the previous one has passed beyond its highest point.

Thus our descendants are our most dangerous enemies, whom we cannot overcome, for they will outlive us, and therefore, without fail, will take the power from our enfeebled hands.'

Carl needed to get away. Less than a week after Emma gave birth to their baby daughter he was off on a two-week cycling tour of northern Italy with his friend Wolf Stockmayer, a doctor from Munich. But things did not work out as planned. On their way home they spent a night in Arona on Lake Maggiore, intending to set off along the lakeshore the next morning and on through Tessin as far as Faido, where they planned to take the train back to Zürich. However, that night Jung had a dream which blew him right off course. He found himself in an 'assemblage of distinguished spirits of earlier centuries', the conversation conducted in Latin. Amongst them was a gentleman with a long, curly wig who addressed him and asked a difficult question. Carl understood the question but did not have sufficient command of the language to answer in Latin. He woke with a start, feeling 'such intense inferiority feelings about the unanswered question that I immediately took the train home in order to get back to work. It would have been impossible for me to continue the bicycle trip and lose another three days. I had to work, to find the answer.' It was the same old thing: a dream which took him back to his boyhood and those terrible feelings. Work was the only solution. Work and home. Emma must have been surprised when he arrived back three days early. Or maybe not. These days she knew how much Carl needed her for his emotional stability. Probably better than he did.

* * *

Earlier that year, in March 1910, Emma had taken a more direct role in Carl's work. The Second International Psychoanalytic Congress was taking place on 30 and 31 March in Nuremberg, but on 8 March, Carl unexpectedly left for America: his old patient Medill McCormick had had a relapse with manic episodes and all the signs of another breakdown. Medill's desperate wife begged Jung to come and attend to him. Carl did not need much persuading, even though the length of the sea passage meant he would be hard pressed to get to Chicago and back in time for the congress.

'*Sehr Geehrter Herr Professor!*' Emma wrote to Freud that same day from Kusnacht, 'I am writing to you in the name of my husband, who suddenly left for Chicago today, where his former patient, McCormick, is seriously ill.' She assured Freud that Jung would be at Nuremberg by 29 March, or the 30th at 5 a.m. at the very latest, and might she meanwhile ask whether Freud would mind being the first speaker and what title he wanted for his lecture? All this in her husband's name, exercising the side of herself as she had done all along, assisting Carl as she had assisted her father with his accounts and business correspondence. But the letter carries on, first mentioning a paper on neurosis theory, then passing on a bit of gossip about someone whom Jung had refused attendance at the congress. Emma was going beyond her normal role and involving herself actively. Stepping out. Cautiously. She wrote again on 16 March, thanking the Herr Professor for his kind letter and offer of help, saying that her husband's assistant Dr Honegger was 'deputising with the patients and looking after the Nuremberg business with me, otherwise I would be rather nervous about everything turning out all right'.

'Now don't get cross with me for my pranks!' Jung wrote to Freud the next day from the Grand Hotel Terminus in Paris. 'You will already have heard from my wife that I am on my way to America. *I have arranged everything so as to be back in time for Nuremberg.* Everything else is so arranged that it will function automatically, i.e. with the help of my wife and the assistance of Honegger, to whom I have entrusted my patients.' Honegger was Carl's most promising student and assistant, so the patients were deemed to be in good hands. But Freud was put in a fluster by all this uncertainty. 'Bleuler is not coming either, and Jung is in America, so that I am trembling about his return,' he wrote to Oscar Pfister, another of the Zürich group who could not attend. 'What will happen if my Zürichers desert me?' But Jung came in time, having rescued Medill once again from his self-destructive urges, chasing him all the way to Chattanooga in Tennessee and back. And the congress was a success too, founding the International Psychoanalytic Association, a key step towards establishing the pre-eminence of the Freudian school of psychoanalysis. At a private meeting of the Viennese group of analysts at the congress, Freud announced: 'Most of you are Jews, and therefore incompetent to win friends for the new teaching. Jews must be content with the modest role of preparing the ground . . . The Swiss will save us – will save me, and all of you as well.'

Emma had now met Freud twice in Vienna and once in Kusnacht. She liked and admired him and sensed he was an ally who understood the dilemmas she faced in her marriage. She wanted to keep in touch with the Herr Professor on her own behalf. So when he answered her letter about the Nuremberg arrangements she wrote back, confiding in him the fears she had had about her husband's enthusiasm for America. For a while she had worried Carl might want to emigrate and, good Swiss woman that she was, the thought horrified her. But now she realised 'America no longer has the same attraction for him as before, and this has taken a stone from my heart. It is just enough to satisfy [his] desire for travel and adventure, but no more than that.' And indeed, Jung wrote to Freud that when he got home he was pleased to find 'wife, children, and house in good shape, and work aplenty'.

As for Emma, having once tasted the pleasures of connecting with the exciting new world of psychoanalysis, she was not about to give it up. When preparations were being made for the next congress, at Weimar the following autumn, she was determined to be part of it.

From now on Emma stepped in for her husband every time he went away, and he seemed to be away more and more. In April he was in Berlin, then home for two days before going to Stuttgart for the annual meeting of the German Society for Psychiatry, then 'working like mad' until July, when he set off on a two-week sailing holiday on the Lake of Constance, the Bodensee, sending his boat ahead; then three weeks' military service, then a trip to London for a consultation at the beginning of September, and so on into 1911, 'gadding about like mad' as he admitted to Freud. When Honegger left to work at another asylum, causing one of Carl's furious outbursts – 'How could he do this to me?' – Emma stepped in again.

* * *

Respite came with the annual spring holiday. This time they took a sixteen-day motoring tour to the South of France, complete with chauffeur, staying in the grandest hotels. These holidays were always the best of times for Emma and Carl, sharing their love of travel, history and books, just the two of them.

At the beginning of July a heatwave hit the Continent that lasted right through the summer and into September. In Zürich people walked in the shade of the chestnut trees, the ladies in white ankle-length cotton dresses, white shoes and stockings, wide straw hats and parasols, the men in waistcoats and shirt sleeves with the open-necked Schiller collars normally only worn to the bowling alley. The heat changed the whole atmosphere of the city: the gardens of the Wirtschaften always crowded, waitresses in long skirts and starched white aprons carrying jugs of Hürlimann beer from table to table, and the lidos on the lake full to bursting. Apparently the supply of lemonade ran out at the annual Cantonal Gymnastics Fest at Winterthur. In the evenings people promenaded along the *quais* until the light faded, listening to the military bands in the pavilions.

Still, the heat did not stop some antimilitarists demonstrating in front of the Italian embassy against the war with Turkey, a few communists and anarchists telling the crowd it was an imperialist war, Italy helped by the British colonialist pirates and French finance. 'Italy has her right for a place in the sun like anyone else,' shouted back Benito Mussolini, the fiery young editor of *Avanti!*, who was in the crowd. The city police had to be called in to calm things down.

On the morning of 16 September, Jung took the paddle steamer from Kusnacht to the Bellvue landing stage and then the tram to Central Station, to meet Freud who was arriving from Bolzano, where he had been spending his annual summer holiday with Martha and the family. Jung had invited him a day earlier but Freud answered that this was 'unfulfillable, and has been for twenty-five years': it was the date of his and Martha's wedding anniversary. In the end Freud stayed with the Jungs for four days before they all set off for the Weimar Congress. Emma, in happy anticipation of Freud's visit, made sure the room on the second floor with a view of the lake was beautifully prepared for their honoured guest: fine bedlinen, a table for work, flowers for decoration, and plenty of books.

The days in Zürich were packed with activity, as Ernest Jones, joining them from Toronto, recalled: 'There were of course seminars, visitors and receptions, so it was by no means a pure holiday. Putnam who was staying in Zürich not Kusnacht, participated in all these activities.' Apparently Freud found time in the packed schedule to give Putnam, their new friend from America, six hours of analysis. He also found time to spend a morning alone with Emma whilst Carl was busy with other matters. They talked mostly about family, his and hers. It is clear from letters written later that Emma confided in Freud about some of the dilemmas in her life, seeing in the Herr Professor something of a father figure, perhaps a replacement for her own father who had failed her so badly. Freud in turn confided more to Emma about his private life than he had to almost anyone else, admitting that his relationship with Martha was 'amortised' as he put it – surely in response to Emma's revelations about the difficulties in her own marriage. But it also reveals how relaxed Freud felt in the presence of this 'charming, clever and ambitious' young woman, as he later described her to Jung.

Charming? Clever? Ambitious? Modest Emma did not see herself in those terms.

Conversation continued as they travelled together to Weimar, Emma included for the first time in the exalted company of Freud, Ernest Jones and James Putnam, though once Carl was present she soon slipped into the background. He was telling them, with his usual energy and relish, about a pedagogic congress in Brussels he had attended, where he spoke on the psychoanalysis of children, a provocative subject at the best of times. When he delivered his bombshell of theories, a few people had left the hall in protest muttering, '*Oh, c'est un homme odieux!*' – an odious man – and '*Vous avez déchaîné un orage*' – you've let loose a storm. It could not have been better. After the Brussels congress, and in triumphant mood, he'd collected Emma from Kusnacht and they'd set off on a mountain tour of the Bernese Oberland. And now here they were speeding through the German countryside at twenty miles an hour, the train belching out steam, on their way to Weimar. To Emma, it seemed, life was looking up.

The formal photograph taken of the attendees at the Third International Psychoanalytic Congress in September 1911, on the steps of the Erbprinz Hotel in Weimar, marks an important moment in the history of the early psychoanalytic movement – and a key moment in Emma's life. There she sits in the front row looking elegant in a dark two-piece suit with a high collar and long skirts – remarkably trim in spite of her four pregnancies – hands in lap, composed, hair nicely but not extravagantly coiffed, a small smile on her lips as she looks straight into the camera, attractive and alert with the pleasure of being part of the company. Beside her, along the front row, sit the other women, the independent ones Emma must until recently have envied. They are nearly all from the Zürich group, Vienna lagging far behind as far as accepting women into the fold. On the far left of the picture sits Maria Moltzer, a nursing sister from the Burghölzli and another of Carl's female admirers; two down is Lou Andreas-Salomé from Germany, sporting a flamboyant fur stole in spite of the warm weather; next to her, with Emma on her other side, sits Beatrice Hinkle from New York in a smart American two-piece with striped collar and cuffs; a lady from Berlin in an enormous hat sits on Emma's other side, flanked by Antonia Wolff, a young woman who had until recently been Carl's patient, looking rather cross and unsure in a girlish dress and hair done up in rolls. Lastly, Martha Boeddinghaus, another of Carl's admirers. Noticeably absent was Fräulein Spielrein, who had taken her final exams in January, passing them with top honours; she hadn't been able to make up her mind whether to attend or not, feeling that Herr Doktor Jung had not given her any real encouragement to do so. A relief no doubt to Emma.

'This time the feminine element will have conspicuous representatives from Zürich,' Jung had written to Freud on 29 August 1911: 'Sister Moltzer, Dr Hinkle-Eastwick (an American charmer), Frl Spielrein (!) then a new discovery of mine, Frl Antonia Wolff, a remarkable intellect with an excellent feeling for religion and philosophy, and last but not least my wife.' The 'last but not least' was written in English, everything else in German.

'Dear friend,' Freud replied. 'We Viennese have nothing to compare with the charming ladies you are bringing from Zürich,' adding: 'I am glad to release you as well as your dear wife, well known to me as a solver of riddles, from the darkness by informing you that my work in these last few weeks has dealt with the same theme as yours, to wit, the origin of religion.' The reference to Emma as solver of riddles is praise indeed from the old wizard of psychoanalysis, and references a comment Jung made about Freud's enigmatic way of expressing himself: 'together with my wife I have tried to unriddle your words'. But clever Freud was surely alluding to this talent of Emma's in a more general sense as well, knowing how much riddle-solving life with Jung involved.

Behind the women seated in the front row of the photograph, and all around them, stand most of the male representatives of the Freudian school of psychoanalysis as it was in 1911: Professor Eugen Bleuler, Ludwig Binswanger and Franz Riklin, Alphonse Maeder, Adolf Keller, Oscar Pfister, all from Zürich; then Otto Rank, Wilhelm Stekel, Paul Federn and Isidor Sadger from Vienna; Sandor Ferenczi from Budapest, Ernest Jones from Toronto, originally of Wales, Max Eitingon, Karl Abraham, Leonard Seif and one or two others from Germany; and finally the Americans: Abraham Brill and James J. Putnam. Everyone dressed in dark three-piece suits with stiff collars, cravat or tie and gold watch chains, over forty attendees, all gathered round the two principal participants: Herr Professor Freud and Herr Doktor Jung. The photograph is artfully arranged, an early example of media management: Jung, the only one in a stylish floppy bow tie, standing behind his wife Emma, leaning forward, his arms resting on the back of her chair, thereby minimising his great bulk and height; Freud next to him, formal in tails, standing a good deal higher thanks to a stool, commanding the company with his presence, making Ferenczi on his right, who was the same height, look tiny by comparison.

Emma spent the two days of the congress listening with keen interest to Professor Putnam of Boston 'On the Significance of Philosophy for the Further Development of Psychoanalysis', Professor Bleuler of Zürich 'On the Theory of Autism', Dr Sadger of Vienna 'On Masturbation', Dr Abraham of Berlin on 'The Psychosexual Foundation of States of Depression and Exaltation', Sandor Ferenczi of Budapest 'On Homosexuality', Professor Freud of Vienna on 'Postscript to the Analysis of Schreiber', her own husband on 'Contributions on Symbolism', and many more. She knew most of the speakers, either from the Burghölzli, or from their visits to Seestrasse. When Professor Putnam visited from Boston, for instance, Jung had picked him up from his Zürich hotel and brought him back to the house for a meal and to meet Emma and the children. To think she might have been married to a Schaffhausen businessman, taking after-noon *kaffee kuchen* with their wives and entertaining dull busi-ness colleagues in the evenings.

A bonus for Emma was meeting Beatrice Hinkle, the 'American charmer'. She was exactly the kind of woman Emma liked. Seated side by side in the front row of the photograph, it is easy to see why: each had a quiet composed elegance – two intelligent women in the company of some of the most advanced thinkers of the age. Emma found she had much to learn from Hinkle, who was an early feminist. Like Emma she came from a wealthy family, but unlike Emma she had been allowed to further her education, studying medicine at the Cooper Medical College in San Francisco. In 1892 she married attorney Walter S. Hinkle and by 1899 she had gained her doctorate with a thesis entitled 'Enuresis in Children'. In 1905 she was appointed San Francisco's city physician, the first woman in the United States to hold a public health position. Later, moving to New York, she founded the first therapeutic clinic in America. By 1909 she was travelling to Vienna to study under Freud, later moving to Jung because she found Freud's rigid sexual theories – male-dominated in her opinion – hard to accept. And now she was here in Weimar, taking an active part in all the discussions and, during the recesses, talking advanced ideas about the place of women in society. What with the congress and the feminist ideas, Emma came home dizzy with a new confidence.

After Weimar, Jung immediately set off for St Gallen and three weeks' military service and Emma took the children to stay with her mother in Schaffhausen. On 30 October, in her new confidence, she did an extraordinary thing. She wrote to Herr Professor Freud – whilst Carl was away and without telling him. During the four days of Freud's visit to Kusnacht, and beneath the friendly collaboration of Weimar, she had detected a certain unease between the two men. Carl had previously sent Freud the first part of a paper he was writing, later to be published as *Transformations and Symbols of the Libido*, and he was waiting anxiously for Freud's reaction, knowing that the ideas expressed in it would not please him.

'Dear Professor Freud,' Emma began:

I really don't know how I am summoning the courage to write you this letter, but am certain it is not from presumption; rather I am following the voice of my unconscious, which I have so often found was right and which I hope will not lead me astray this time. Since your visit I have been tormented by the idea that your relation with my husband is not altogether what it should be, and since it definitely ought not to be like this I want to do whatever is in my power. I do not know whether I am deceiving myself when I think you are somehow not quite in agreement with 'Transformations of the Libido'. You didn't speak of it at all and yet I think it would do you both so much good if you got down to a thorough discussion of it.

It is an astonishing letter, courageous, and perhaps a little rash. Whatever would Carl say if he found out? The letter shows how much Emma has absorbed of the language and methods of psychoanalysis. She talks of 'following the voice of my unconscious', having picked up an incipient rift between the pair. Emma the peacemaker, Emma the defender of her husband.

'Or is it something else?' she intuits, perhaps referring to Freud's admission that his own marriage was 'amortised', as well as his concerns about his children:

If so, please tell me what, dear Herr Professor; for I cannot bear to see you so resigned and I even believe your resignation relates not only to your real children (it made a quite special impression on me when you spoke of it) but also to your spiritual sons; otherwise you would have so little need to be resigned. Please do not take my action as officiousness and do not count me among the women who, you once told me, always spoil your friendships. My husband naturally knows nothing of this letter and I beg you not to hold him responsible for it or let any kind of unpleasant effects it may have on you glance off on him. I hope nevertheless that you will not be angry with your very admiring
Emma Jung.

On the same day Carl sent Freud a postcard from his military barracks in St Gallen. The last ten days had completely worn him out, he said, having suddenly been detailed for a mountain exercise in the back of beyond where he was quite out of touch with the rest of humanity. He was returning home from these brutalities the next day. 'I am glad you are home again and no longer playing soldiers, which is after all a silly occupation,' replied Freud, never mentioning Emma's letter. Nor the fact that he had answered it.

'My dear Professor Freud,' Emma wrote back, presumably hiding away in her room so Carl did not find out what she was up to. Freud's letter is lost, but it is not hard to work out what it contained. 'Your nice kind letter has relieved me of anxious doubts, for I was afraid that in the end I had done something stupid. Now I am naturally very glad and thank you with all my heart for your friendly reception of my letter, and particularly for the goodwill you show to all of us.' Freud must have remonstrated that he was not really resigned, as she put it, nor defensive, but Emma was not minded to accept it: he had not even let them sympathise with his toothache, she pointed out, versed as she now was in repression, projection, resistances, the unconscious and so forth. She also explained why she had mentioned 'Symbols': because she knew how eagerly Carl was waiting for the Herr Professor's opinion, having often said he was sure 'you would not approve of it', and was thus 'waiting with some trepidation'. She reckoned it was a residue of Carl's father (or mother) complex, which was probably being resolved in the book, 'for actually Carl, if he holds something to be right, would have no need to worry about anyone else's opinion'. Not content to let sleeping dogs lie, she reminded Freud of the conversation on that first morning when he told her about his family: 'You said then that your marriage had long been "amortised", now there was nothing more to do except – die. And the children were growing up and then they become a real worry, and yet this is the only true joy.' This comment of his had seemed so significant to her, she had to think of it again and again. She fancied it was intended for her private information, and not for Carl, because Herr Professor's comments likely referred 'symbolically' to her husband at the same time.

She moved boldly on, oblivious to the likely reaction of the venerable old man who was, after all, the founder of the Viennese school of psychoanalysis. This was so unlike the normally reserved Emma, so alien, it suggests the heady experience of Weimar had temporarily sent her off beam: 'I wanted to ask then if you are sure that your children would not be helped by analysis. One certainly cannot be the child of a great man with impunity, considering the trouble one has in getting away from ordinary fathers. And when this distinguished father also has a streak of paternalism in him, as you yourself said!'

On she went, in full swing: 'You said you didn't have time to analyse your children's dreams because you had to earn money so that they could go on dreaming. Do you think this attitude is right? I would prefer to think one *should not* dream at all, one should live.' This was an extraordinary statement to make to the father of psychoanalysis. Then she returned to the subject of her own life: 'I have found with Carl also that the imperative "earn money" is only an evasion of something else to which he has resistances.' How hard she is trying to understand it, to explain it all to herself. 'Please forgive me this candour, it may strike you as brazen; but it disturbs my image of you because I somehow cannot bring it into harmony with the other side of your nature, and this matters so much to me.' The Protestant Swiss young woman might have misunderstood the Jewish Viennese professor's quirky comments about his children, but she was spot on about her own Swiss husband.

'You may imagine how overjoyed and honoured I am by the confidence you have in Carl, but it almost seems to me as though you were sometimes giving too much – do you not see in him the follower and fulfiller more than you need?' she carried on, quite oblivious that she was going too far. 'Doesn't one often give much because one wants to keep much? Why are you thinking of giving up already instead of enjoying your well-earned fame and success? Perhaps for fear of letting the right moment for it pass you by? Surely this will never happen to *you*. After all, you are not so old that you could speak now of the "way of regression", what with all these splendid and fruitful ideas you have in your head! Besides, the man who has discovered the living fountain of ps.A.[psychoanalysis] (or don't you believe it is one?) will not grow old so quickly.'

She signed off with a final flourish: 'And do not think of Carl with a father's feeling: "he will grow, but I must dwindle", but rather as one human being thinks of another, who like you has his own law to fulfil. Don't be angry with me. With warm love and veneration, Emma Jung.'

When Freud wrote to Jung on 12 November he made a point of praising *Transformations and Symbols of the Libido*, but added: 'Sometimes I have a feeling that [your] horizon has been too narrowed by Christianity.' Again Freud never mentioned Emma's letters, only sent a 'heartfelt to your wife and children'. But he'd had enough.

'You were really annoyed by my letter, weren't you?' Emma wrote back to Freud on 14 November:

I was too, and now I am cured of my megalomania and am wondering why the devil the unconscious had to make you, of all people, the victim of this madness. And here I must confess, very reluctantly, that you are right: my last letter, especially the tone of it, was really directed to the father-imago, which should of course be faced without fear. This thought never entered my head; I thought that, knowing the transference side of my father-attitude towards you, it would all be quite clear and do me no harm. After I had thought so long before writing to you and had, so I believed, fully under-stood my own motives, the unconscious has now played another trick on me, with particular finesse: for you can imagine how delighted I am to have made a fool of myself in front of you. I can only pray and hope that your judgement will not prove too severe.

Once again Freud's letter is lost but Emma's response makes it quite clear how sharp it was. There was one thing she neverthe-less wanted to vigorously defend herself against: the way the Herr Professor reacted to her 'amiable carpings', as he put it. She had not suggested Carl should set no store by Freud's opin-ion, she protested, it went without saying that one recognised authority. It was just that her husband had been so anxious and uncertain, which had seemed superfluous to her. But now she saw that she had got it wrong: Carl had not been worried about what Freud would say, he had just been analysing his attitude to his work and discovered some resistances to it. The worry about

Freud was really only a pretext for not going on with the self-analysis. She realised now that she had projected something 'from my immediate neighbourhood into distant Vienna and am vexed that it is always the nearest thing that one sees worst'.

'You have also completely misunderstood my admittedly uncalled-for meddling in your family affairs,' she went on, showing that she was quite up to defending herself:

> Truthfully I didn't mean to cast a shadow on your children. I know they have turned out well and have never doubted it in the least. I hope you don't seriously believe that I wanted to say they were 'doomed to be degenerate'. I have written nothing that could even remotely mean anything of the sort. I know that with your children it is a matter of physical illness, but just wanted to raise the question whether these physical symptoms might not be somehow psychically conditioned, so that there might for instance be a reduced power of resistance. Since I have made some very astonishing discoveries in myself in this respect and do not consider myself excessively degenerate or markedly hysterical, I thought similar phenomena possible with other people too. I shall be grateful for enlightenment.

This is Emma at her best: lucid, honest, open, and quite assertive. She continues:

> That you think it worthwhile to discuss your most personal affairs with me is something for which I thank you with all my heart. What you tell me sounds so convincing that I simply have to believe it, although much in me struggles

against it. But I must admit you have the experience and I do not, consequently I am unable to make any convincing rejoinders. You are quite right about one thing, though: despite everything and everybody, the whole affair is only a blessing in clumsy disguise which I beg you to forgive. Please write nothing of this to Carl; things are going badly enough with me as it is.

Emma Jung.

No one could have guessed that things were going 'badly enough' for Emma. So generous towards her husband, there was no sign of the struggle going on within. Not until she had to explain herself to Freud.

Freud wrote back quickly and sympathetically. 'Heartfelt thanks for your letter,' she replied on 24 November 1911:

Please don't worry. I am not always as despondent as I was in my last letter. I was afraid you were angry with me or had a bad opinion of me; that was what made me so downhearted, especially because my main complex was hit. Usually I am quite at one with my fate and see very well how lucky I am, but from time to time I am tormented by the conflict about how I am to hold my own against Carl. I find I have no friends, all the people who associate with us really only want to see Carl, except for a few boring and to me quite uninteresting persons. Naturally the women are all in love with him, and with the men I am instantly cordoned off as the wife of the father or friend.

Here was Emma's dilemma in a nutshell, expressed with courageous honesty – or, more accurately, her many dilemmas, all connected to the central one: Carl.

Yet I have a strong need for people and Carl too says I should stop concentrating on him and the children, but what on earth am I to do? What with my strong tendency to autoeroticism it is very difficult, but also objectively it is difficult because I can never compete with Carl. In order to emphasise this I usually have to talk extra stupidly when in company. I do my best to get transferences and if they don't turn out as I wished I am always very depressed. You will now understand why I felt so bad at the thought that I had lost your favour, and I was also afraid Carl might notice something. At any rate he now knows about the exchange of letters, as he was astonished to see one of your letters addressed to me; but I have revealed only a little of their content. Will you advise me, dear Herr Professor, and if necessary dress me down a bit? I am ever so grateful to you for your sympathy.

With warmest greetings to you and yours, Emma Jung

Imagine Carl's surprise when he found a letter in Freud's hand on the hall table one morning – addressed to his wife, not himself. Emma's private correspondence with Herr Professor Freud ceased forthwith.

10

A Difficult Year

1912 was a difficult year in the Jung house. A crisis was looming, provoked by Carl's split from Freud. Emma had sensed it coming, but she had no idea it would be so bad, no idea it would bring her husband close to a complete emotional breakdown.

It was a difficult year for Zürich too, plagued by a series of strikes which disrupted the usually peaceful and orderly way people went about their business. The *Neue Zürcher Zeitung* reported that the strike at the automobile factory at Schlieren was quickly followed by one at the Firma Gauger in the Unterstrasse. The city police intervened to protect the strike-breakers, good family men as the newspaper described them, who were beaten up as they went to work. Finally, the local military were called in. One striker was fatally wounded in the fray and his funeral on 24 April became the occasion for a noisy demonstration. The strike leaders accused the military of being the 'lapdog of Capitalism'. Activists and revolutionaries, some from Russia and Germany, told demonstrators to 'Stop work! It's the only way the Capitalists will recognise that they belong to you, not you to them!' People marched through the streets, banners waving, distributing flyers: 'Down with the military! Swiss soldiers, don't shoot your brothers! Long live the International Revolution!' There were 60,000 foreign workers in Zürich: Germans, Austrians, Czechs, Italians, most anti-militarists. The younger generation from the Yiddish-speaking Jewish community joined the protests whilst their elders worked day and night in small factories making buttons and ribbons and trimmings for the clothing industry. There were instances of sabotage and in the 'red' working-class district of Ausser Sihl, a bomb-making factory was found, stocked with tin boxes, nitroglycerine and fuses.

Neither Emma nor Carl was especially political, but this was shocking. The class war appeared to be entering new territory: it was not just the anarchists and revolutionaries creating the unrest now, but regular social democrats too. It was all the odder since the financial pages of the newspapers showed that everything else seemed to be going so well: millions invested in stocks and shares, post office traffic of letters, telegrams, telephone calls all growing exponentially. A glance through the newspaper advertisements showed you could buy everything in Zürich: the new Waterman fountain pen from America, beer from Bavaria, Indian teas directly from Fortnum & Mason department store in London. Over 3,000 Swiss owned an automobile, the majority living in Zürich. Trade was thriving thanks to the efficiency of the new railway system, with wheat arriving from Russia and machinery of all sorts from America. Locally the watch industry, the chocolate factories, the embroidery business, the tourist and hotel trade – all thriving. Add to this the recognised fact that Switzerland had the finest sanatoriums and offered a humane refuge to those fleeing political repression and religious intolerance, and you have Helvetia, proud and independent. The burghers had their rights, could read and write; they were not beaten, or hanged; they elected their administrators, and could marry whom they wished. It was amusing to Carl to find that a new word had entered the vocabulary: psychoanalysis, though not everyone knew what it meant, or even how to spell it.

In January 1912, an unusually mild month with clear blue skies, the citizens of Zürich had woken up to psychoanalysis. Carl was obliged to go public and defend it in the pages of the *Neue Zürcher Zeitung*, writing to Freud that there were 'mighty rumblings in Zürich over Psycho-Analysis. The Keplerbund is sponsoring a public lecture against this abomination. Protest meetings are afoot!' In the evening edition of 5 January, amongst articles about the invention of the electric iron, central heating, the typewriter and the latest craze for winter sports – which was 'only for Englishmen and women and our rich who like to imitate them' – came a sharp attack on this 'Psycho-Analysis', which pretended to be scientific but was nothing more than dubious hypotheses. Rumours spread about strange goings-on and loose morals. 'Zürich is seething,' Jung wrote. 'Psycho-Analysis the talk of the town. One can see here how worked up people can get . . . These people are vermin that shun the light.' He needed a break. He went to Lugano on his own for a week at the end of March. Then Emma joined him and they toured northern Italy together.

When Emma took the paddle steamer into town to meet Marguerite that summer, she found a city in buoyant mood: ladies' fashions more modern, corsets less restrictive, the skirts shorter, showing an ankle. Some ladies even rode bicycles now, braving the traffic and the danger of the wheels getting stuck in the tramlines. As ever, the sisters would make for the fashion house Grieder on the Bahnhofstrasse, with its wondrous variety of silks and fine cottons, embroidery, lace and ribbons laid out for them on wide wooden counters by deferential assistants. Further down the street were the new travel agents advertising exotic destinations. Marguerite and Ernst were keen travellers and sometimes Emma joined them on short trips when Carl was too busy or too distracted to get away. After shopping, the sisters might sit on the terrace of the Hotel Baur-au-Lac or if the weather was inclement, in one of the fashionable cafés – the Sprüngli or the Wiener – taking coffee with a slice of Schwartzwalder Torte or millefeuilles, two young women in elegant day wear, hats and gloves, talking about their children and their husbands.

Equally they might browse the newspapers and talk about the worsening international situation: there was trouble in the Balkans, France wanted Alsace Lorraine back, the Germans wanted a bigger fleet to match the British, the socialists wanted to be rid of the monarchies. One in five of all Zürich workers were German, and some warned that social discontent still lay not far below the surface. On 12 July there was a general strike. The City Fathers blamed those 'other Germans': the activists, socialists and anarchists from Berlin who came to do mischief in Zürich because they could not get away with it in the Reich. Shopkeepers pulled down their blinds again as strikers ran through the streets waving flags and shouting slogans through megaphones en route to the Volkshaus, where a large demonstration was taking place. Smallholders from Forch coming down the hill in their wooden carts for the weekly market were told to go back whence they came. The city police had to be reinforced by the canton military again, Battalions 62, 64 and 67 standing guard in front of city hall, rifles at the ready. But you could see their hearts were not in it, most of them being the sons of local farmers or factory workers. Several arrests were made, including Fritz Brupbacher, the proto-communist doctor and known associate of Russian 'agents provocateurs'. By 13 July the strike was officially over, but on the 15th, after another demonstration outside the Tonhalle, over 700 workers were fired or had their wages docked. Some German members of the workers' unions were deported under emergency legislation passed on 18 July, implemented with immediate effect.

Then everything went back to normal. The *Neue Zürcher Zeitung* reported that on a sunny Sunday a large crowd gathered in Sihl Tal to watch a flying machine display by the glamorous pilot Maffei, who liked to sport his leather cap back to front. Faces turned skywards as he circled above before landing bumpily on the grass. The organisers had to rope off an area round the plane to stop spectators crowding in as Maffei posed beside it, smiling for the cameras. Other photographs showed fathers and small sons gazing at the plane in wonderment whilst mothers and daughters sat on rugs further off. Stalls selling bratwurst were doing brisk business, and there was beer for the men, apple juice or lemonade for the women and children. Later in the afternoon Maffei climbed back into his flying machine, reversed his cap, his assistant turned the propellers, and the flying machine bumped and hopped a few feet along the grass before rising unsteadily in the air, turning in the direction of Leimbach, the red and white Swiss flag fluttering gaily behind.

In September there was an event of far greater significance: the state visit by Kaiser Wilhelm II. Besides the growing

German population of Zürich, there were plenty of burghers who saw the visit as historic and an honour, possibly including Carl, who had been so excited to catch a glimpse of the Emperor Franz Joseph in Vienna. Others were indifferent, some actively opposed. Whilst grand ladies ordered their day and evening gowns from their favourite fashion houses and practised their curtsies and their best High German, others muttered dissent. The old Swiss song 'Wir Brauchen Keine Schwaben in Der Schweiz' ('We Don't Need Any Germans in Switzerland') was whistled by workers in the streets. The anarchists were crueller, chanting: *'Alle Raeder stehen still, wenn mein starker Arm es will'* ('All the wheels will stop turning if my strong arm wants it'), a reference to the Kaiser's withered left arm. The Kaiser's days were filled with receptions, presentations and entertainments, all amply covered by the newspapers. But everyone knew that his real interest was the military man-oeuvres and his inspections of the Swiss Army. What no one knew, not even the Swiss President Herr Doktor Ludwig Forrer, was that in the same month, September 1912, the Kaiser had held his first *Kriegsrat*, war council, in preparation for a war which he and his advisers already considered inevitable. Nor did they know that in London, King George V's advisers were doing the same.

There were pictures of President Forrer and selected members of the Swiss government in top hats and tails, standing on the red-carpeted platform of Zürich Central Station awaiting the imperial train, with an honour guard in full dress uniform and a military band at the ready. As the state carriage proceeded down Bahnhofstrasse, outriders clattering alongside, the Kaiser apparently told President Forrer how pleased he was to be in Zürich, a city close to his heart, and congratulated him on his iron-fisted suppression of the strikers. To which Forrer responded by reminding the Kaiser that unfortunately there were many Germans amongst the red agitators. 'Quite so! Quite so!' replied the Kaiser, rather missing the point.

By November there were more demonstrations, by the unemployed of Zürich at the Volkshaus, and later that month by the ordinary *Hausfraus* – hatless, wrapped in black woollen shawls. Emma felt for them: the price of everything had gone up, but not their husbands' wages: eggs were 11 rappen each now, milk up by 2 rappen, cheese up by 10 rappen, stewing beef 2 francs 10 rappen a kilo, which was more than their men earned in a day. Bread up, flour up, oats, briquettes for the stove, sugar, and rents as well. The weather report that day predicted rain. Black umbrellas went up as the crowd dispersed.

At 228 Seestrasse, Kusnacht, half an hour along Lake Zürich by paddle steamer, the strikes, the Kaiser's state visit, the sinking of the *Titanic* in April, revolution in China and unrest in Persia, all remained no more than a backdrop to the dramas taking place in Carl and Emma's own lives – private dramas, hidden away from public view. It was like an island, that house, during 1912, surrounded by the garden with its long yew-lined drive, lawns, terraces, rose bower, boat house and summer pavilion looking out across the lake, cut off from the rest of the world, introverted, battling with itself, all the optimism of the previous year and the Weimar Congress forgotten.

The problem lay with Carl. Split Carl. Personality No. 2 was back in the ascendant, bringing him close to an emotional breakdown. Day after day he locked himself away in his *Cabinet*, writing, delving into his unconscious, dicing with madness. He called it his experiment. 'It is of course ironical that I, as a psychiatrist, should at almost every step of my experiment have run into the same psychic material, which is the stuff of psychosis and is found in the insane,' he wrote later.

The crisis was brought about, as Emma had divined, by his imminent break with Freud following the publication of his *Transformations and Symbols of the Libido*, written in two parts during 1911 and 1912 and presenting Jung's own highly personal theory about the libido. 'When working on my book about the libido and approaching the end of the chapter "The Sacrifice" I knew in advance that its publication would cost me my friendship with Freud,' Jung recollected in the tranquillity of old age, years after the drama of the falling-out had subsided into psychoanalytic history. 'For I planned to set down in it my own conception of incest, the decisive transformation of the concept of libido, and various other ideas in which I differed from Freud.' He was turning Freud's theory of the libido and the oedipal complex of the child's repressed incestuous wish towards the parent on its head. 'To me incest signified a personal complication only in the rarest cases,' Jung wrote:

> Usually incest has a highly religious aspect, for which reason the incest theme plays a decisive part in almost all cosmogonies and in numerous myths. But Freud clung to the literal interpretation of it and could not grasp the spiritual significance of incest as a symbol. I knew that he would never be able to accept any of my ideas on this subject. I spoke to my wife about this, and told her my fears. She attempted to reas-

sure me, for she thought Freud would magnanimously raise no objections, although he might not accept my views.

Emma was wrong. Carl was right. Freud was profoundly angry and felt betrayed. He had deceived himself on one point and one point only, as he wrote to Sandor Ferenczi on 26 November 1912: that Jung was 'a born leader', where in fact he was 'immature himself and in need of supervision'.

Emma had seen it coming. She had been 'tormented by the idea' that Carl and Freud's relationship was in trouble when she had written in secret to Freud on 30 October the previous year. She sensed Freud was not in agreement with 'Transformations of the Libido': 'You didn't speak of it at all and yet I think it would do you both so much good if you got down to a thorough discussion of it.' She was not deceiving herself, but she was on the wrong track if she thought Freud would be prepared to discuss it. Carl knew Freud would not compromise. He knew it, but he did not know the break would send him almost insane.

Now Carl was assailed by all the old horrors from his unconscious. Without Emma keeping the steady rhythm of family life going he might have cracked. 'The unconscious contents could have driven me out of my wits,' he wrote. 'But for my family, and the knowledge: I have a medical diploma from a Swiss university, I must help my patients, I have a wife and [five] children, I live at 228 Seestrasse in Kusnacht – these were the actualities which made demands upon me and proved to me again and again that I really existed, that I was not a blank page whirling about in the winds of the spirit, like Nietzsche.' Here he was again, as he had been in the garden of his father's parsonage, staring at the stone jutting out from the wall, asking himself: am

I the stone or is the stone me? 'Whenever I thought that I was the stone, the conflict ceased.'

There had been early signs of friction with Freud, when Jung began to take his first steps on the path to his own theory about the libido. In June 1910, Jung had written to Freud about a talk he had given on 'Symbolism' to a group of Swiss psychiatrists in Appenzell – the most traditional of Swiss cantons, still deeply embroiled in the superstitions and occult of Alpine people cut off from the more modern ways of the towns. His lecture had been received with enthusiasm, he told Freud, and he promised to send him a copy. Apart from the logical, he posited, there was another language based on symbols and images. Logical think-ing was thinking in words. Analogical or fantasy thinking was emotionally toned, pictorial and wordless, a rumination on materials belonging to the past – archaic and unconscious. Jung spent every evening immersed in 'the overflowing delights of mythology'.

'I read your essay with pleasure the day it arrived,' replied Freud, taking his time, promising to respond to it soon. His response, when it arrived, caused dismay in the Jung household. The letter only exists in copy form, the original being lost or destroyed. Freud suggested Jung's use of the word 'symbolical' was vague and not really correct when formulated in such general terms, and noted that Jung had 'wholly overlooked' the customary symbolism of the wedding ceremonial, wreaths, rings, and so on. But what Freud really disliked was the sentence: 'Sexuality destroys itself', which, he said, had provoked 'a vigor-ous shaking of the head'. Further, he did not like Jung's idea that modern dreams were a residue of the ancients: 'This would be more apt if the ancients, who lived in mythology, had not also had dreams.' As to myths, they were originally psychological, later overlaid by adaption to the calendar and thence projected into the realm of natural phenomena. In other words, there was not much in the essay Freud could agree with. 'As usual, I have mentioned only objections and made no comments on many things I liked very much,' he added by way of encouragement, but ending with the comment: 'Despite all its beauty, I think, the essay lacks ultimate clarity.'

It was the beginning of the parting of the ways, though neither man could yet admit it. Freud was for the knowable, based on scientific 'proof'; Jung was increasingly for the unknowable, where no final 'proof' could exist. In his reply to Freud he admitted the essay was just a 'rough sketch', but he held fast to the basics, so fast that it had to be expressed in italics: 'What it boils down to is a *conflict at the heart of sexuality itself*. The only possible reason for this conflict seems to be the *incest prohibition* which struck at the root of primitive sexuality.' Freud, still hoping his brilliant colleague would become his heir, answered with careful conciliation: he saw now that his criticism was quite premature but he warned that proof was nevertheless needed. He threw in another of his Wilhelm Busch humorous couplets, by way of apology: 'The ass is stupid, thence his name/The elephant is not to blame.'

Jung had managed to write Part One of *Transformations and Symbols of the Libido* without too much difficulty, but Part Two confronted him with appalling problems, plunging him into the nightmare of his own fantasies – pages and pages of rambling, incoherent material, embracing parapsychology, spiritualism, the occult and his new investigations into archaic symbols and rituals and myths. 'It seemed to me I was living in an insane asylum of my own making. I went about with all these fantastic figures: centaurs, nymphs, satyrs, gods and goddesses, as though they were patients and I was analysing them.' And astrology, for without astrology how could one understand mythology? 'I make horoscope calculations in order to find a clue to the core of psychological truth,' he wrote to Freud. He was like a detective, hunting for clues. He remembered the Burghölzli patient Emile Schweizer, the schizophrenic who thought he was God and that the sun had a penis. Was this not also archaic symbolism and myth? He remembered his assistant Honegger, who had worked on Schweizer's fantasies. After leaving the Burghölzli, Honegger had committed suicide, taking an overdose of morphine. Jung saw too late that Honegger had himself been suffering from incipient schizophrenia. He re-examined the theory of the death wish which he had first discovered with Fräulein Spielrein. Round and round he went, lost in his own labyrinth behind the doors of his *Cabinet*. 'I have tried to put the "symbolic" on a psychogenetic foundation, i.e., to show that in individual fantasy the *primum movens*, the individual conflict, material or form (whichever you prefer) is mythic, or mythologically typical,' he wrote to Freud, still trying to explain.

But that was all over now. Effectively, he was plunging into his own unconscious fantasies, back to the exalted and terrifying images of his childhood and adolescent dreams and visions. To his first biographer, Barbara Hannah, he admitted that he was in a deep depression the whole time he was writing *Transformations and Symbols of the Libido*: 'While working on the book I was haunted by bad dreams,' he admitted. 'It took me a long time to see that a painter could paint a picture and think the matter ended there and had nothing whatever to do with himself. And in the same way it took me several years to see that it, the *Psychology of the Unconscious* [i.e. *Transformations and Symbols*], can be taken as myself and that an analysis of it leads inevitably into an analysis of my own unconscious processes.' It went against everything he had learnt during his medical training about the need for scientific objectivity, but he could not stop. He heard voices dictating to him, and he had to obey. 'They would have torn me up, or I could have been split off,' he admitted years later. 'I had known that everything was at stake and that I had to take a stand for my convictions. I realised that the chapter 'The Sacrifice" meant my own sacrifice.'

Why was the break with Freud so bad? Emma thought she knew: she had touched on it in her letter to Freud of 6 November 1911. She thought Carl was so anxious waiting for the Herr Professor's opinion on 'Symbols' because it was a residue of the father (or mother) complex which was probably being resolved in the book. 'So perhaps it is all to the good that you did not react at once so as not to reinforce this father–son relationship,' she added, revealing how much she had learnt about psycho-analysis in general and her husband in particular over the years. Nor did Carl disagree, at least not in the early days. In 1909 he had been so grateful for Freud's support in the Fräulein Spielrein crisis that he had to keep telling himself he would have done the same for a friend: 'I had to tell myself this because my father-complex kept on insinuating that you would not take it as you did but would give me a dressing down more or less disguised in the mantle of brotherly love.' A year later he was still able to admit why he struggled to write to Freud: 'The reason for my resistance is my father-complex, my inability to come up to expectations (one's own work is garbage, says the devil) . . .'

Now he also recalled his first year at grammar school, when he had his vision of God sitting high above Basel cathedral and the wicked thought, the 'enormous turd falls upon the sparkling new roof, shatters it, and breaks the walls of the cathedral asunder', leaving him with 'an enormous, indescribable relief', so that he wept for happiness and gratitude. It was then that it dawned on him for the first time that God could also be something terrible, not only good, and split Carl embraced the idea: 'I am a devil or a swine,' he thought. 'I am infinitely depraved.' He never spoke about it, nor the phallus dream, nor his carved manikin, but kept everything hidden away. 'My entire youth can be understood in terms of this secret.' It was his great achievement, he felt, not to disclose the secret to anyone, but it induced 'an almost unendurable loneliness'. In time he would tell Emma. Finally, when he was sixty-five, he wrote it down in the 'Protocols' which became the basis of *Memories, Dreams, Reflections*.

Jung held his secret so close that few people noticed any difference in him, seeing only Personality No. 1. He was always busy, always on the move. By August he was away on military service again, quickly followed by another of his trips to America, leaving Emma to placate his patients and see to all his correspondence as usual. He did not get back until the beginning of November. Away from home he was released from his demons. His letters to Emma are full of vitality, with no sign of Personality No. 2. First he gave a series of lectures in New York on 'The Theory of Psychoanalysis'. This covered many of the same ideas as *Transformations and Symbols*, but the style was quite different, clear and lucid with no ramblings or confusions. Speaking with impressive fluency in English, albeit with a strong Swiss accent, he pinpointed the differences between his own and Freud's approach fairly, describing Freud's concept of the libido as 'an exclusively sexual need' whereas he saw it 'in the more general sense of passionate desire'. It was only in his lecture on the Oedipus complex that he described Freud's definition as 'too narrowly restrictive'. He also gave two-hour-long seminars every day, addressing scores of psychiatrists and neurologists, and clinical lectures at Bellevue Hospital and the New York Psychiatric Institute on Ward Island, and he spoke at the New York Academy of Medicine. After New York he travelled to Chicago to check on his old patient Medill McCormick and treat some new ones, many of whom later became his patients in Kusnacht. Then he travelled on to Baltimore and thence to Washington where he met Medill's cousin, Theodore Roosevelt, but also visited the other district of Washington – black Washington, where the poverty was in shocking contrast to rich white Washington. He was keen to treat and analyse the 'Negro' patients in a local asylum, which he did, using them as research material in much the same way as he had used the inmates of the Burghölzli asylum.

Throughout his tour of America, Personality No. 1 prevailed, with Carl exuding energy, charm and charisma. To catch a glimpse of Personality No. 2 you have to look elsewhere, to the black-and-white studio portrait photograph which accompanied an article and long interview in the *New York Times* on Sunday 29 September. Here Jung looks very different: eyes hard and dark behind his gold-rimmed spectacles, mouth shut tight beneath the small moustache; no smiles, no charm, and none of his usual robust humour.

Carl Jung, New York, 1912.

'America Facing its Most Tragic Moment' ran the alarming headline, followed by a series of equally alarming thoughts about the current state of America, a country Jung had visited three times, none for more than a few weeks. The interviewer set out Dr Jung of Switzerland's theory of the unconscious: 'He believes that, if a man can understand his hidden motives and impulses, he comes into a new power.' The word 'unconscious' signified 'all the facts we refuse to face'. Jung's classrooms and clinics were constantly crowded, wrote the reporter, baffling other doctors.

Jung was well on the way to becoming a *succès fou*, and this was only increased by the combative nature of his comments. 'When I see so much refinement and sentiment as I see in America I look always for an equal amount of brutality. The pair of opposites – you find them everywhere,' he opened, firing on all cylinders. He found a lot of prudery in the United States, he wrote, which was always the cover for brutality. This is what America needed to address. 'It seems to me you are about to discover yourselves.' Listing the country's past achievements he noted that in the present era America did not appear to know it was in danger. A choice had to be made. Americans had built their big cities with their theatres and clubs and cathedrals, all ready and waiting to be used to some great end once they discovered themselves. They should stop hiding from themselves. Then their success in the big achievements of art and literature would astound Europe as their big business and philanthropy did now. Conquerors descend to the level of those they conquer, he suggested. 'I notice that your Southerners speak with the Negro accent,' he continued, apparently oblivious to overstepping the line, and then, right over the top now: 'your women are coming to walk more and more like the Negro.'

Allowing for some deliberate exaggeration for effect, this was still wild and provocative, crazy stuff. Carl was letting Personality No. 2 freewheel, and once he was in his stride there was no stopping him. Had Emma been with him it might have been different. But she wasn't. 'In America,' he preached:

your women rule their homes because the men have not yet learned to love them, it takes much vital energy to be in love . . . The American husband is very indignant when he comes to me for treatment . . . and I tell him it's because he is brutal on one hand and prudish on the other . . . You think your young girls marry European husbands because they are ambitious for titles . . . I say, they like the way European men make love, and they like to feel we are a little dangerous. They are not happy with their American husbands because they are not afraid of them.'

Jung himself later said of this period of his life: 'The journey from cloud-cuckoo-land back to reality lasted a long time . . . Pilgrim's Progress.'

At home in Kusnacht, Emma waited for Carl's return. 'Dear Professor Freud,' she wrote on 10 September 1912, unaware of how grave the situation between her husband and Freud would shortly become. 'The offprints of Part II of "Transformations and Symbols" have just come out and you must be the first to receive one.' She went on with 'heartfelt sympathy' to share the worry and the hope that Freud's daughter Mathilde, who had been very ill, would soon recover. 'We had a dismal summer too, the children had whooping cough and now measles; Carl was away nearly all summer; since Saturday he has been on to America after spending only one day here between military service and departure. I have so much to do now that I don't let too much libido travel after him to America, it might so easily get lost on the way. Please greet all your dear ones from me and give my best wishes to your daughter. With kindest regards, Emma Jung.' The letter was full of warmth, reaching out to Freud and his family. She also confided in him about her 'libido', which she says she has to protect. Libido was still a term in transition then, everyone using it for their own purposes. For Emma it sounds more like 'out of sight, out of mind': that is, do not think too much about what Carl was up to in America. Reserve your energy for your own life.

Once Carl was back from America all Emma's diplomacy was undone. Carl turned against Freud with a vengeance, unable to come to terms with his rejection of *Transformations and Symbols of the Libido* – that is, Freud's rejection of Carl's very self. He saw through his little tricks, he wrote in a furious letter on 18 December, using the French word *truc*: 'You go around sniffing out all the symptomatic actions in your vicinity, thus reducing everyone to the level of sons and daughters who blushingly admit the existence of their faults. Meanwhile you remain on top as the father, sitting pretty. For sheer obsequiousness nobody dares to pluck the prophet by the beard and inquire for once what you would say to a patient with a tendency to analyse the analyst instead of himself. You would certainly ask him: "*Who's* got the neurosis?"'

Freud knew the answer. 'As regards Jung,' he wrote to Ernest Jones, 'he seems all out of his wits, he is behaving quite crazy. After some tender letters he wrote me one of utter insolence . . .' To Ferenczi he wrote that Jung was 'behaving like a florid fool and a brutal fellow, which he certainly is'. The love affair between Freud and Jung was well and truly over.

* * *

Around Christmas, Carl had another significant dream. He was in a magnificent Italian loggia, high in the tower of a castle, with pillars, marble floor and balustrade, sitting on a gold Renaissance chair in front of a table of rare beauty made of an emerald-like green stone. His children were sitting at the table too. Suddenly a white bird like a dove descended and came to rest on the table. He signed to the children to be still so as not to frighten the bird away, whereupon it transformed into a little girl, about eight years old, with golden blonde hair. She ran off with the children to play amongst the colonnades of the castle. When she came back she put her arms tenderly round his neck, then transformed back into the dove, which then spoke to him in a human voice: 'Only in the first hours of the night can I transform myself into a human being, while the male dove is busy with the twelve dead.' Then it flew off into the blue air and Carl awoke.

What could it signify, this dream? He thought of the twelve apostles, the twelve months of the year, the signs of the zodiac, but he found no solution. More dreams followed, all echoing the kind of dreams he had had when he was a boy, taking him back through the layers of time, to ancient castles and tombs and burial vaults, down, down, deeper and deeper into his unconscious. But they were unable to relieve his conscious mind, which remained disorientated 'as if under constant inner pressure. At times this became so strong that I suspected there was some psychic disturbance in myself.' Finally, he decided the only solution was to submit himself consciously to his own unconscious.

First, a memory from his eleventh year surfaced, the time when he had gone through an extreme crisis and started to build houses and castles with gates and vaults, 'passionately' at first with building blocks, then outside in the garden with stones and mud. 'To my astonishment, this memory was accompanied by a good deal of emotion.' He realised he had to re-establish contact with that period and that self. He had to overcome endless resistances but finally he resigned himself: he went out into the garden and started looking for suitable stones along the lake shore and in the shallow water, building 'cottages, a castle, whole villages, a church. But the altar was missing.' He didn't know what to do about that, till, one day, walking along the shore, he caught sight of a red stone, a four-sided pyramid about an inch and a half high. 'I placed it in the middle under the dome, and as I did so, I recalled the underground phallus of my childhood dream. This connexion gave me a feeling of satisfaction.'

Every day now, after taking his midday meal with Emma and the children in the *Stube* with its long windows looking out over the lake, Carl would go out into the garden, weather permitting, to build and play like the child he once was, till the first of his afternoon patients arrived. And he went back to it every evening, day after day. He asked himself: 'Now, really, what are you about? You are building a small town, and doing it as if it were a rite!' There was no answer to the question, only an inner certainty that he was discovering his own myth. The activity released a stream of fantasies which he later wrote down and illustrated, much like a medieval manuscript. From then on, every time he came to a blank wall in his life, he painted a picture or hewed a stone. 'Father would be down there fitting rocks together,' Franz remembered. 'He was a genius at that. He would build towers and houses and churches until he had whole villages. I would cut reeds for the roof beams and fill the little houses with sand so they wouldn't fall down.'

'My father says he chose. I do not believe he chose,' said Franz. 'I believe he had no choice. Can you imagine what it must be to think that you might be going mad? That you might fall for ever into the void?' It was a crisis which his father could no longer avoid. 'For years after he and Freud parted, my father could do no work,' said Franz. 'He placed a gun in his night-stand, and said that when he could bear it no longer he would shoot himself. Other people fell away, and he was alone . . . Think of my mother. Think of her. Can you imagine living with a man who slept with a gun by his bed and painted pictures of circles all day?'

What indeed was Emma to make of it? In 1912 she was thirty, married for nine years, the mother of four young children and the wife of a man who locked himself away in his *Cabinet* for hours on end, or played in the garden, building stone and mud villages, castles and churches. How could she tell anyone about it? Who would understand? In the old days she could have confided in Freud, but no longer.

It was all the more puzzling because outwardly so much carried on as normal. Patients came and went, the children played in the garden alongside their father, the housemaids went about their work cleaning and dusting, the cook cooked, the gardener gardened. Emma oversaw it all, giving everyone their daily duties, helping the children with their homework, and disciplining them, because Carl had no wish to do so, had no wish, really, to act as a father at all. But she also carved out time for herself, reading and writing at her desk by the window over-looking the lake, continuing her own researches. Because, as

Freud noted, Emma was not only charming, but clever and ambitious too.

All the while Carl's Personality No. 1 functioned alongside Personality No. 2 with no outward sign of conflict. 'It was most essential for me to have a normal life in the real world as a counterpoise to that inner world. My family and my profession remained the base to which I could always return, assuring me that I was an actually existing, ordinary person,' he wrote, again and again. Years later he acknowledged it was only Emma and the children who had got him through this period of semi-madness after the split with Freud: 'My family and my profession always remained a joyful reality and a guarantee that I also had a normal existence.' By his own account it was not until the end of the First World War that he emerged from the darkness and confusion.

11

Ménage à Trois

Besides her husband's fragile mental state, Emma had another recurring problem. As early as January 1910, just seven years into their marriage, Carl had written to Freud airily declaring he had decided the prerequisite of a good marriage was the licence to be unfaithful. There must have been a row with Emma about it, because she 'staged a number of jealous scenes, groundlessly', as Carl had put it. 'At first my objectivity got out of joint (rule 1 of psychoanalysis: principles of Freudian psychology apply to everyone except the analyser) but afterwards snapped back again, whereupon my wife also straightened herself out brilliantly.' The two sentences follow on, one from the other, but there appears to be no connexion between the two in Carl's mind. Emma, usually so contained, beside herself with jealousy, humiliated, making 'scenes'; Carl denying everything, protesting his innocence. It was happening all over again.

Polygamy had been on Jung's mind for some time. At first he could only envy cheerful polygamists such as Max Eitingon, the doctor at the Burghölzli, or Otto Gross, Jung's one-time patient with the wife who let him do pretty much as he pleased. 'I've never really had a mistress and am the most innocent of spouses,' as he'd written to Freud in March 1909. But Emma already had her suspicions because in another letter to Freud in the same month Jung admitted: 'the devil can use even the best of things for the fabrication of filth. Meanwhile I have learnt an unspeakable amount of marital wisdom, for until now I had a totally inadequate idea of my polygamous components despite all self-analysis.' He knew how to lay that devil now, he wrote, but he had been 'churned up hellishly inside'. Through all the drama and pain, his relationship with his wife had 'gained enormously in assurance and depth'.

But, as Emma knew, the problem was not only Carl. Women seemed to find Herr Doktor Jung irresistible. 'I really cannot remember whether I have already told you that Sister Moltzer is reproaching herself for having painted too black a picture of Fräulein Boeddinghaus,' a disingenuous Jung wrote to Freud in September 1910. 'Between the two ladies there is naturally a loving jealousy over me.' Naturally. Both women were volunteer nurses at the Burghölzli, where Jung still treated some patients. Sister Moltzer assisted Jung in his work while training to be a psychoanalyst herself.

Jung knew that many men in his circle kept mistresses, most of them in secret. In addition to Eitingon and Gross, there was Ernest Jones, who ended up being blackmailed, Ferenczi, who even had an affair with the daughter of the woman who was his mistress, and later Oscar Pfister, one of the Zürich Protestant pastors connected to the Burghölzli. Often, in those early days of the psychoanalytic movement, the mistresses were patients or former patients. Only Freud seems to have steered more or less clear, in spite of the gossip about his unmarried sister-in-law, Minna. To his American friend, James Putnam, Freud explained in 1915: 'I stand for an infinitely freer sexual life, although I myself have made very little use of such freedom.' And to Jung: 'My Indian summer of eroticism that we spoke of on our trip has withered lamentably under the pressure of work. I am resigned to being old and no longer even think continually of growing old.'

An underlying problem was the fear of pregnancy. 'All is well with us, except for the worry (another false alarm, fortunately) about the blessing of too many children,' Jung had written to Freud in May 1911. 'One tries every conceivable trick to stem the tide of these little blessings, but without much confidence. One scrapes along, one might say, from one menstruation to the next. The life of civilised man certainly does have its quaint side.' He didn't need to tell Freud, who, after fathering six children with Martha, had decided the only solution was abstinence. A subject rarely discussed was prostitution. The old town of Zürich was teeming with prostitutes, the sordid underbelly of a clean, prosperous, upright society. And God's punishment? The scourge of syphilis. If Carl watched his father crack up and die an early death from an unnamed torment, Emma watched her father die in terrible pain from syphilis, blind, shunned and shamed. It bound them deeply together, Emma and Carl, the horror of their two fathers' deaths.

By 1912 rumours were rife that Jung was having an affair with the volunteer nurse Maria Moltzer, who had come to the Burghölzli to get away from her family in Holland, rebelling against her father, the powerful proprietor of the Bols gin family business. Even Freud in Vienna had heard the rumours. Members of Emma's family and close friends have said that in the course of their long marriage Emma threatened divorce three times. The Maria Moltzer 'infatuation' may have been one time, not because Emma wanted to end her marriage, but to give Carl a fright – bring him into line – because deep down she knew he loved her. The last thing Carl wanted was divorce. Whenever Emma threatened to leave him he fell ill with stomach cramps, influenza, bouts of depression. And whenever one of the women started making demands, seriously threatening their marriage, he fell into a panic. Infatuation was Carl's problem, not discontent with his marriage. 'Except for moments of infatuation my affection is lasting and reliable,' he explained to Freud.

But now Emma was confronted not only with the Maria Moltzer affair. There was another relationship too, one quite different from the rest. With the sure instinct of a worried wife Emma sensed this was much more than the usual infatuation. Jung, already teetering on madness, seemed to be losing his wits.

The woman in question was Antonia Wolff, the same Fräulein Wolff who had only recently attended the Weimar Congress, with apparently no threat to Emma. There she sits in the front row of the famous photograph, a thin, intense, frowning, unconfident young woman, two seats away from Emma, who is smiling, elegant, more confident now that the threat of Sabina Spielrein has passed, with Maria Moltzer four places to her right and Martha Boeddinghaus – that other admirer of Carl's – three to her left. And behind them stands Carl, bending solicitously over Emma as though to say: 'You are mine, and I am yours.'

'A new discovery of mine,' was how Carl had described Wolff to Freud in his letter to finalise the Weimar conference arrangements: 'a remarkable intellect with an excellent feeling for religion and philosophy'. Knowing the significance of Jung's shift away from the 'provable' science of psychoanalysis to the unprovable, transcendental realms of myth and religion, and knowing Jung's weakness for such women, Freud no doubt foresaw the dangers. But not Emma. Not yet. Unsure of herself, so serious she 'never laughed', and not obviously attractive, Fräulein Wolff was probably the last of those women in the front row Emma felt she had cause to worry about.

Antonia – Toni – Wolff had started out as one of Carl's patients. Her mother had brought her to see Jung in the summer of 1910, on the recommendation of friends, and having consulted various other doctors about her profoundly depressed daughter. The problem had begun some six months earlier after the death of Toni's father on Christmas Eve 1909, probably from cancer. Toni was twenty-one, the eldest of three daughters, and no one knew why she had fallen into such a deep, inconsolable state. The family lived on the Zürich Berg, the exclusive enclave where many of Jung's admiring *Pelzmäntel* ladies lived. The Wolffs were leading members of old wealthy Zürich, Herr Wolff having made a great deal of money importing silks from the Far East, where he had lived for many years. When he finally came back to Switzerland he was in his early forties and soon married Toni's mother, twenty-year-old Anna Elizabeth Sutz. Their villa on Freiestrasse was filled with Japanese antiques and beautiful paintings and a fine library.

Toni and her sisters, Erna and Susi, had a childhood in Zürich as privileged as Emma Rauschenbach's in Schaffhausen. Toni was academic, highly intelligent, interested in philosophy, mythology and religion, but emotionally fragile, even as a girl.

Erna wanted nothing more than a good marriage, a good house and children; and Susi was unconventional, temperamental and artistic. Toni was her father's favourite, permitted to spend evenings with him in his *Herrenzimmer*, the room in fine villas reserved exclusively for the men. Like Emma, Toni had been prevented from continuing her studies at university in favour of the traditional path of a good marriage. To that purpose her father sent her off to one of Switzerland's many housekeeping schools for young ladies, and for some further months in England. But this was not to interfere with the main task. 'A woman who does not have the household completely under control and in all details is a failed creature; yet her whole life does not have to revolve around it,' Herr Wolff wrote to his daughter by way of loving explanation. 'It is self-evident that in addition to this you are free to pursue more serious intellectual and aesthetic studies, but not to the exclusion of practical ones.'

By the time her mother brought her to Herr Doktor Jung at 228 Seestrasse, Toni Wolff's depression was entrenched. The chasm left by her father's death could not be filled and her unofficial studies in philosophy and theology at the university, which is all her father had finally permitted, had led to nothing substantial. She was lost and disturbed. Even so, the extent of it was hard to understand. What was it about her favoured relationship with her father that had so derailed her? At first Jung made no headway. But once they talked mythology and symbolism, the subject currently obsessing Jung, embroiled in writing *Transformations and Symbols of the Libido*, they found their common ground and her depression began to lift. To Jung's pleasure Fräulein Wolff was also interested in astrology, and they spent many hours drawing up astrological charts. Carl

Meier, a young doctor who later left the Burghölzli to become Jung's assistant, described Jung at this time as being 'very near psychosis' himself, his crisis resembling 'a schizophrenic episode'. Jung used to say Toni Wolff was the one case of schizophrenia he had been able to cure.

Unlike Sabina Spielrein, who had been an official patient at the Burghölzli asylum, Toni Wolff was Jung's private patient so there are no proper case notes to record her treatment, only a general sense that by September 1911 she was essentially 'cured' – and well enough to attend the Weimar Congress. By then the treatment had moved to its next stage, just as Eugen Bleuler at the Burghölzli always advised, and the patient was set to work: in this case going to Zürich Central Library to help Jung with his research. Not long after the congress Jung had written to Freud about his ill-humour: 'I was furious because of something that happened in my working arrangements. But I won't bother you with that . . .' Then in a later letter: 'With us everything is peaceful and serene and my wife is working conscientiously at etymology.' There are many references to etymology in *Transformations and Symbols of the Libido*. Seeing the dangers, Emma must have insisted on helping Carl with the research herself.

Toni Wolff was a strange, other-worldly being. 'Intense' is the word most often used to describe her. She rarely laughed, hardly ever smiled, though when she did it lit up her face. Her sister Susi, the lively one who married Hans Trüb, another of the Zürich group of psychoanalysts, described Toni as 'never totally in life', and never coming fully alive except through Jung. 'Toni was all spirit. It was almost as if she *had* no body,' Franz Jung recalled. 'A ghost-like figure, haughty and forbidding' was how another contemporary described her. 'She had very changeable looks, sometimes beautiful and sometimes quite plain,' said another: 'Her extraordinarily brilliant eyes – mystic's eyes – were always expressive.' These days Toni dressed more elegantly, in the wealthy Swiss bourgeois style: hats and gloves, expensive shoes, a long black cigarette-holder in her hand because she was an inveterate smoker. She lived with her mother at the family villa on Freiestrasse until her mother died in 1940, at which point she moved into an apartment in her sister Erna's house, taking her personal maid with her. She had few real friends, and if at all, she was the kind of woman who preferred men. 'If you were a woman she had to lord it over you. If you were a man it was peaches and cream.' Many found her unapproachable, but she had a great gift. 'In her presence, inner pressure became images,' said Tina Keller, another of the Zürich group around Carl and Emma. 'She helped Jung see his images and talk with them.'

The relationship with Fräulein Wolff started at a time of crisis for Jung, battling with the loss of Sigmund Freud, his father figure, stirring up all the old feelings about his own father: 'After the parting of ways with Freud, a period of inner uncertainty began for me. It would be no exaggeration to call it a state of disorientation. I felt totally suspended in mid-air, for I had not yet found my own footing.' He started to develop a new method with his patients, not guiding them with psychoanalytic theory but waiting to see what they would tell him of their own accord – a conversation between two equals, sitting facing one another; the kind of approach that Freud, seated formally in a chair behind the analysand's couch, preserving professional objectivity, would disapprove of. Just at this jagged point in his life, entered Toni with her gift of seeing images and talking to them.

'Please write nothing of this to Carl,' as Emma had said to Freud, without mentioning Fräulein Wolff, and more likely thinking of Maria Moltzer.

At home, Carl's odd behaviour continued. He was frequently so overwrought he had to do yoga exercises to hold his emotions in check. At the midday meal with Emma and the children he might be completely silent and distracted, lost in his own thoughts, and the children had to sit quietly and stop fidgeting. Equally he might talk his head off, enchanting them with his stories, throwing his head back, laughing 'down to his shoes', as Franz put it. Then it was back to his *Cabinet* again. Or out into the garden, walking up and down, talking away to Philemon, one of the mythological figures from his unconscious who had first appeared to him in a dream representing superior insight. 'He [Philemon] was a mysterious figure to me. At times he seemed to me quite real, as if he were a living personality. I went walking up and down the garden with him, and to me he was what Indians call a guru,' Carl wrote later. But the guru is a separate person from the seeker of wisdom, whereas Philemon was all Carl.

By autumn 1913 Jung feared he had a psychic disturbance. He knew there had been some mental illness in earlier generations of his family, and that his mother heard voices and had to go 'away' when he was a child. 'Then I came to this. "Perhaps my unconscious is forming a personality that is not me, but which is insisting on coming through to expression." I don't know why exactly, but I knew to a certainty that the voice that had said my writing was art had come from a woman . . . Well I said very emphatically to this voice that what I was doing was not art, and I felt a great resistance grow up within me . . .' If it was merely art, then it was not symbolic. Years later he wrote: 'I knew for a certainty that the voice had come from a woman, a talented psychopath who had a strong transference to me.' It was Maria Moltzer's.

In September there was another Psychoanalytic Congress, this time in Munich. Jung's talk was about 'Psychological Types' and the idea that people could be divided into extroverts or introverts, terms which were quite new at that time. It was an important, imaginative insight – Carl at his best. But his emotional state was not good. 'Two years ago there was robust gaiety and exuberant vitality in Jung's booming laughter,' said one attendee, 'but his seriousness now is made up of pure aggression, ambition and intellectual brutality.' Freud and Jung talked politely, but it was all over between them.

A month later Jung was on the train to join Emma and the children for a weekend in Schaffhausen, to celebrate his mother-in-law Bertha's fifty-seventh birthday, when he had what he called an overpowering vision: a monstrous flood covered all the land between the North Sea and the Alps. He realised it represented a frightful catastrophe, the whole sea turning to blood and thousands of bodies drowned. The Alps of Switzerland grew higher and higher to avoid the flood. What could it mean? In October he wrote to Freud, finally breaking off relations completely and resigning his position as editor of the *Jahrbuch*. He also started to write his secret diary again, the one he had abandoned when he became engaged to Emma. His last entry then had been: 'I am no longer alone with myself, and I can only artificially recall the scary and beautiful feeling of solitude. This is the shadow side of the fortune of love.'

In November 1913 Ernest Jones reported to Freud from America that an acquaintance of theirs 'considers Jung to be mentally disordered'. Jung saw it differently:

When I had the vision of the flood in October of that year 1913, it happened at a time that was significant for me as a man. At that time, in the fortieth year of my life, I had achieved everything that I had wished for myself. I had achieved honour, power, wealth, knowledge, and every human happiness. Then my desire for the increase of these trappings ceased, the desire ebbed and horror came over me. The vision of the flood seized me and I felt the spirit of the depths, but I did not understand him. Yet he drove me with unbearable inner longing and I said: My soul, where are you? Do you hear me? I speak, I call you – are you there? I have returned. I am here again.

This is the beginning of chapter one of Jung's visionary *Liber Novus – The Red Book* as it is now known – a work of extraordinary creative imagination which he began at this time of crisis and continued to work on for the next sixteen years, leaving it unfinished. The theme of the book is how Jung regains his soul which he felt he had lost during the years since giving up his secret diary and his marriage to Emma. It takes the form of conversations with various biblical, mythological and symbolic figures Jung encounters along the way. First he wrote the encounters in a series of notebooks bound in black leather, the 'Black Books'; later he transcribed them into a massive red leather-bound tome, 189 pages of Old Italic script, sometimes Latin, mostly German, accompanied by brilliantly coloured manuscript illuminations, the translation of his emotions into symbolic images, in the style of a medieval Bible. The first initial of the first page, the 'D' of *Der Weg des Kommenden*, 'The Way of What Is to Come', is quite beautiful, reminding one that the landscape paintings Jung did in Paris as a young man already had an otherworldliness to them. The 'D' is red, a snake wearing a crown rising from a black cauldron of fire on the stem. It is set in a Swiss landscape, a lake with a sailing boat, a green fish and plants in the foreground, village with church spire in the middle ground, chain of Alps in the background, the whole dominated by a luminous blue. The text is in Latin, from Luther's Bible: 'Isaiah said: Who hath believed our report? And to whom is the arm of the Lord revealed?'

When Jung finally put the *Liber Novus* to one side it lay undisturbed in his *Cabinet* for the next twenty years, and would not be published until 2009. It comes from the period Jung described as 'the most important time in my life. Everything else is to be derived from this.' The only people he shared these inner experiences with at the time were a few 'close associates'. And Emma, and Toni Wolff.

On 12 December 1913 Jung was sitting alone at his desk in his *Cabinet* in the dim light of the stained-glass windows depicting the Crucifixion and Passion of Christ, 'thinking over his fears', and he resolved to again let himself drop, deep down into his unconscious. 'It was as though the ground literally gave way beneath my feet, and I plunged down into dark depths. I could not fend off a feeling of panic.' He landed in a soft, sticky mass, in complete darkness, but with a feeling of relief at having landed. Once his eyes got used to the gloom, a deep twilight, he saw the entrance to a dark cave, guarded by a dwarf with leathery, mummified skin:

> I squeezed past him through the narrow entrance and waded knee deep through icy water to the other end of the cave where, on a projecting rock, I saw a glowing red crystal. I grasped the stone, lifted it, and discovered a hollow underneath. At first I could make nothing out, but then I saw that there was running water. In it a corpse floated by, a youth with blond hair and a wound in the head. He was followed by a gigantic black scarab and then by a red, newborn sun, rising up out of the depths of the water.

He tried to replace the stone, but a fluid welled out. It was blood. 'A thick jet of it leaped up, and I felt nauseated. It seemed to me the blood continued to spurt for an unendurably long time. At last it ceased and the vision came to an end.' Dark caves and projecting rocks and precious stones – just like the dreams and visions of his childhood.

He realised it was a hero-myth, a drama of death and renewal. Six days later he had a dream: he was with an unknown brown-skinned man, a savage, in a lonely rocky landscape. It was before dawn and the eastern sky was bright, the stars fading. 'Then I heard Siegfried's horn sounding over the mountains and I knew we had to kill him.' Siegfried appeared high on the crest of a mountain, the first rays of the rising sun behind him. He drove with furious speed down the precipitous slope, in a chariot made of the bones of the dead. 'When he turned the corner, we shot at him, and he plunged down, struck dead.' Now Carl, the dreamer, fled, filled with disgust and remorse at having killed something so great and beautiful, and frightened that the murder might be discovered. But then there was a tremendous downfall of rain and he knew it would wipe out all the traces. 'I had escaped the danger of discovery; life could go on, but an unendurable feeling of guilt remained.' As he tried to go to sleep the voice spoke again: 'If you do not understand the dream you must shoot yourself,' it said. He thought of his military service revolver in the drawer of his night table, loaded, and felt very frightened.

Emma, understanding more about Carl's complex personality now, and fearing for his emotional stability, saw to it that everyday life at Seestrasse carried on as normal. The children, Aggi aged ten, Gret eight, Franz six and Marianne (known as Nannerl) aged three, still noticed nothing unusual: their father saw patients, as many as seven a day five days a week, he gave lectures, kept up a lively correspondence, went on his travels and Alpine hikes, took friends sailing on Lake Zürich, and every day if the weather was fine he took coffee with their mother on the terrace after the midday meal, often with visitors, because there were always visitors. The only difference was that Fräulein Wolff seemed to be around more and more. The children were told to call her Tante Toni and be nice to her, and to include her in family life. They resented it deeply and couldn't understand it. They didn't know that a year earlier their father had had the dream which caused him to embark on a relationship with a woman he had known for three years – the dream of the dove that turned into a little girl with golden hair – and that he had almost let himself drown in despair one day, swimming in the lake, trying to work it all out.

The children didn't like Tante Toni. She wasn't good with children, didn't know how to speak to them, and she didn't care for card games. They took to playing pranks on her when their father was absent. Or their mother. Because although it was an agony for Emma to have to bear what was fast becoming a *ménage à trois*, she never allowed the children to be impolite to Tante Toni. She knew Carl was in a desperate state, despite outward appearances, and she decided that somehow or other she was going to have to come to terms with it.

On 18 March 1914 Emma and Carl's fifth child, Emma Helene, known as Helene, or 'Lil', was born. The baby was another of what Carl called their 'little blessings', but this would be the last. The family say Emma was angry to find herself pregnant again – angry with her fate as a woman, and angry with Carl who did not have to endure the pain of labour or the restrictions of motherhood. Only two weeks after the birth at home in Seestrasse, Carl left for Ravenna on a bicycle tour with his friend Hans Schmid. Emma and the baby went to stay with her mother at Schaffhausen, where she was the unhappy recipient of cheerful postcards telling her all about the beauty of the Ravenna mosaics. Carl's mother and Trudi moved into the house to help the children's maid look after the other four. When Carl returned he and Emma moved into separate bedrooms. There were to be no more children. Emma was thirty-two.

That spring, Jung embarked on a flurry of resignations and withdrawals. On 20 April 1914 he resigned as president of the International Psychoanalytic Association, along with fifteen other Swiss members, all rejecting 'the papal policies of the

Viennese', to which Freud reacted, writing to Karl Abraham: 'So we are rid of them at last, the brutal holy Jung and his pious parrots.' Ten days later Jung resigned as lecturer in the medical faculty of the University of Zürich, though no great effort was made to persuade him to stay. By 24 July he was in London at the Psycho-Medical Society giving a talk, 'On Psychological Understanding', describing Freud's methodology as reductive, comparing it to someone who tried to understand a Gothic cathedral through its mineralogical aspect. On the 28th he was in Aberdeen addressing the British Medical Association on 'The Importance of the Unconscious in Psychopathology'. Outwardly confident and combative, inwardly fragile, Jung could not stop working. 'As a psychiatrist I became worried, wondering if I was not on the way to "doing a schizophrenia" as we said in the language of those days,' he admitted years later to a friend. 'I was just preparing a lecture on schizophrenia to be delivered at a congress in Aberdeen, and I kept saying to myself: "I'll be speaking of myself! Very likely I'll go mad after reading out this paper."'

Meanwhile the world itself was going mad. On 28 July, Austria declared war on Serbia. The following day the Tsar of Russia, Serbia's ally, mobilised his army along the Austrian border. On 1 August, Germany declared war on Russia. On the 2nd the Tsar signed a declaration of war. Three days later Britain declared war on Germany.

The sequence of events was so quick that many people found themselves in the wrong place and the wrong country. Prince Heinrich of Germany, the Kaiser's younger brother, was visiting his cousin Georgie, George V, at Buckingham Palace. The Dowager Tsarina of Russia was also in England, on her way to

Sandringham with her sister Alexandra, the Dowager Queen of Britain. All over Europe nannies and governesses, tutors, valets and ladies' maids, businessmen and diplomats, husbands and wives, raced for home. In Zürich a French theatre troupe had to cut short their stay. Italian bricklayers downed tools and made their way to Zürich Central Station, along with 42,000 German factory workers, waiters and hotel employees, some singing the patriotic 'Wacht am Rhein' as they marched up the Bahnhofstrasse with the Slavs and Russians, the Poles and Bohemians, carrying their small suitcases and cardboard boxes done up with string, making for the third-class carriages.

Carl wrote a hasty letter to Emma on 1 August: he had heard news that Switzerland was mobilising and had left Scotland immediately in order to get a place on a boat to Holland, since it was more likely he could get a train through Holland and Germany than through France. There was an enormous fleet under steam on the Firth of Forth. If Maria Moltzer was in Holland, he would bring her with him. 'Greet the children from me,' he ended his letter. 'And be kissed by me, my darling, your Carl.'

All the apocalyptic dreams and visions he had been experiencing were not signs of incipient schizophrenia, he decided, but premonitions of war from his subconscious. 'Finally I understood. And when I disembarked in Holland on the next day, nobody was happier than I.'

12

The Great War

Life in Switzerland during the First World War was insular, cut off from the rest of the world with all her borders shut, safe from the surrounding horror. After early fears that Germany would annex this small land it soon became clear that Swiss neutrality suited everyone, not least the Swiss themselves.

No one thought the war would last more than six months, there had been trouble in the Balkans too often before. Meanwhile Zürich flourished as a cultural centre, welcoming refugees from every side. Artists, writers, musicians, revolutionaries, political exiles, spies, agitators all congregated there, foregathering in the cafés and the parks, talking, arguing, drinking, scheming. To the Dada artist Hans Arp, arriving in Zürich from Alsace, these were enticing times. 'In my memory it seems to have been almost idyllic. In those days Zürich was occupied by an army of international Revolutionaries, Reformers, poets, painters, philosophers, politicians and pacifists.' Idyllic and free. The Café Odeon by the Bellevue Platz overlooking Lake Zürich was a favourite meeting place, every table occupied by some expatriate group, each talking their own language. They would sit for hours over little more than a *kaffee creme* or *thé citron*, money being scarce, the place thick with smoke and intrigue.

Stephan Zweig, staying in a Zürich hotel, learnt to take care over what he said or wrote: 'The chamber maid, emptying the waste-paper basket, the telephonist, the waiter who stood suspiciously close and took a bit too long over his service, all working for a foreign power.'

The Dadaists bagged a round table in the window. Opposite sat the writers Wedekind, Leonhard Frank, Franz Werfel and their friends. At another table a group of Russian dancers gathered round the choreographer and soloist Sakharoff, whilst further along the avant-garde painters Baroness Werefkin and Jawlensky entertained their own circle of friends. The art dealer Paul Cassirer, astute promoter of the French Impressionists and Post-Impressionists, came in with his cronies. And in one corner, quite alone and apparently unperturbed by all the noise, sat General Wille, the man who had reformed the Swiss Army in 1912 just in time for the general mobilisation of August 1914, smoking his cigars and reading the *Neue Zürcher Zeitung*, at a table always reserved for him.

The Irish writer James Joyce preferred the more elegant, richly panelled Kronenhalle on the opposite side of the street, where he sat with Ferruccio Busoni, the Italian composer and concert pianist, a pacifist who refused to perform in any of the warring countries, and René Schickele, the French writer and essayist, conversing over wine or champagne. Anarchist performers and writers Hugo Ball and his wife Emmy Hemmings, branded as traitors when they fled Berlin, also preferred it at the Kronenhalle, though they had no money, so Ball took a job playing the piano in a *variété* down in Zürich old town. Ball and Hemmings protested loudly against the war and in favour of international socialism and revolution, founding the

Cabaret Voltaire in the Niederdorfstrasse – 'an International Cabaret', as Ball described it. They opened their doors to a bemused public with an evening of 'Negro music', Tristan Tzara, the Rumanian poet, chanting '*Boum, boum boum – drabafja mo gere drabatja mo bonoooo*', costumed in a cardboard suit of armour, his head covered by a cardboard box, robot-like. Emmy Hemmings appeared from behind a makeshift green curtain, lips bright red, kohl-black eyes, bright green pullover, standing in the spotlight beside the piano singing: 'This is how we live, this is how we live, this is how we die, this is how we murder our Kamaraden in the Dance of Death. Sleep on.' At the end of the evening she sold the song lyrics written on postcards to members of the public, collecting money, and thanking their friends in Russia and France, announcing that Dada stood for 'a rejection of War Fever and a disgust at petty-minded Bourgeois values – a new beginning and a return to the primitive'. The press was scathing. 'We reject this Bolshevism in art, as we do Bolshevism proper,' announced the *Winterthur Tageblatt*. The audience, mostly enthusiastic supporters from the Café Odeon, stamped their feet, shouted, threw orange peel and coins. A few students and passers-by in evening clothes joined in, but the orderly bourgeois of Zürich kept away. They had better things to do than watch *so en Schmarre* – such rubbish.

Vladimir Ilyich Ulyanov – Lenin – was able to enter Switzerland via Bern without even a passport in those early days of 1914, before suspicion of foreigners took hold. All he had to do was mention the highly respected socialist politician Hermann Greulich, who had worked with Marx and Engels on the First International, and he was in. He was not the only one. Soon Switzerland was awash with anarchists and leftists from all the surrounding countries, enthusiastically welcomed by the Swiss socialists but increasingly viewed with alarm by the rest. By September 1915 Robert Grimm, leading member of the Social Democrat Party, was organising a conference at Zimmerwald attended by Lenin and a contingent of communists and pacifists as well as socialists. Their call was for a new International and a reawakening of the class struggle. They were in little doubt that this war, dragging on for over a year now, was a capitalist imperialist war, setting worker against worker, when what they needed was unity to fight the real war: the class war.

Lenin and his wife Nadya stayed in Bern till February 1916 when they moved to Zürich. One of the main reasons was Zürich Central Library, much superior to the one in Bern, and the very same library Emma attended to pursue her studies whenever time and family permitted. Mostly Lenin worked on his thesis 'Imperialism, the Highest Stage of Capitalism', in which he argued that imperialism was the product of capitalism, capitalists being bound to extend their search for profit into territories where labour and raw materials were cheaper, with war the inevitable result. On Sundays when the library was shut, Lenin and his wife liked to walk up the Zürich Berg, finding a spot with a good view of the Alps and the lake, lying in the grass, both reading. First they lived at Geigerstrasse, but it was next to the Ruff sausage factory and the smell was awful. So they moved to Spiegel Gasse, not far from the Cabaret Voltaire, where they remained for the rest of their stay in Zürich, waiting for the call to return to Russia. They preferred the local Sussihof restaurant to the Café Odeon or the Kronenhalle, and Lenin loved his local bowling alley where he became a keen member of the club. The Swiss socialists were generous in their financial support of the couple, but the Lenins found they weren't very revolutionary. In April 1917 they suddenly disappeared, back to Russia and some real revolution.

During the early days of the war everyday life in Zürich went on much as before, only the gas lamps were turned low, making the streets more hazardous as night fell. The trams still ran, ringing their bells at bicyclists and automobilists, the *Droschken* drivers still stood by their stamping horses outside Central Station waiting for custom, the market gardeners still came down from the uplands of Forch on their farm carts every Tuesday and Thursday. The fowl market in Augustiner Platz was still as busy as ever. The commodities agents were quick to adjust to war, selling their stocks in Schneider-Creusot and Standard Oil in favour of General Motors, Baltimore and Ohio and Royal Dutch – American stocks, not European. Neutral Zürich thrived as a commercial and financial centre, with the tiny kiosk of the telegraph office at Central Station busy day and night.

But once it was realised the war would last longer than six months, agents switched to raw materials and foodstuffs, shifting goods out of the warehouses as fast as possible before all the borders were completely shut – the Basel frontier already gone but Geneva still open for the time being – buying sugar and coffee, cotton and coal, flour, rice, raisins, oats, noodles, tins of sardines, cocoa beans, condensed milk. *Hausfraus* who could afford it started to stockpile, dragging the food home with the help of their maids and storing it in the cellar. Shop windows started to empty. The Jelmoli department store ran out of rayon stockings, lampshades and perfumes. There were queues everywhere. The black market boomed. Gold was the answer, not paper money, which was useless.

At the city hall emergency measures were brought in: a strengthening of police powers and a limit to personal freedoms. Street demonstrations were banned. Everyone had to pull together, it was said. *Gasthofs* and hotel restaurants were henceforth to close by 11 p.m. Landsturm Battalions 54 and 57 and 58 stood at the ready in preparation for general mobilisation. New recruits assembled ramshackle in school yards across Zürich in the shade of the chestnut trees; young lads with woollen vests knitted by their mothers under their shirts; farm boys, office workers, shop assistants, tradesmen, apprentices, fitted out in caps and uniforms with red collars, heavy leather belts and sabres, not yet a military unit but getting there – set up not to fight the war, but to patrol the borders, to keep Switzerland safe. Whole divisions marched out of the yards and into the streets, officers on horseback keeping up the pace, flags flying, military bands playing, housewives in aprons waving from their doors, on their way to the barracks at Winterthur.

Carl was hardly back from his travels before he was called up to the Army Medical Corps, taking his rifle out of the cupboard in the front hall at Seestrasse, and donning his old uniform, ready to vaccinate hundreds of new recruits. Later he tended to the sick and occasional wounded. Later still he was deputed to the POW camps set up in neutral Switzerland, serving all sides in the war though mostly the officers, not the men. It was a shock for the good ladies of Zürich gathered in their starched Red Cross uniforms at Central Station, military band playing, waiting for the first train of French wounded from the Battle of the Marne, bunches of alpine roses in hand by way of welcome, to see half-dead soldiers carried off the train on stretchers or limping down the platform on crutches, or led along, eyes bandaged, blind. And these no older than their own sons and brothers and husbands patrolling the Swiss borders.

Military service for Jung never lasted more than a few weeks at a time. He wrote to Emma from one billet or another – describing his work, asking her to visit, send him fresh clothes, and bring him more books as his new work 'Psychological Types' began to take shape – signing off with his usual *Küsst Dich Dein Carl*, 'with kisses from your Carl'. But Emma could rarely visit, what with the demands of a new baby and four other children, and the house to keep going. Unlike Toni Wolff or Maria Moltzer, who could get away easily.

Carl loved Emma. But at the same time he held fast to his ideas on polygamy, assuring her it made no difference to the way he felt about her. As far as Carl was concerned, men and women were different: 'In a man this relationship is not exclusive. When the average man permits comparison of his wife to other women he says, "She is my wife among women." To the woman, though, the object that personifies the world to her is *my* husband, *my* children, in the midst of a relatively uninteresting world.' The man was free to have a family, pursue a career, have affairs; the woman remained confined to the home with little chance of a life outside – views not unusual for the times. Less usual was that Carl conducted his affairs openly, not in secret. Emma, meanwhile, remained monogamous and devoted. 'I think she had many admirers,' a friend of the Jungs later recalled. 'She was very beautiful.' But she would never have had an affair, thought the friend, because of her background and the type of person she was.

Inveterately fair-minded, Emma tried to do right by Toni Wolff, even by Maria Moltzer, but seemed unable to do right by herself, hardly recognising herself any more. In her distress she turned to Carl, as she always did, describing her predicament and her feelings of dissociation. He told her these feelings were a necessary outcome of the development of the personality. He had had them himself, so he understood. But all the promises and understanding in the world could not make up for the fact that she, still young, was now living the life of a celibate, standing by as other women enjoyed her husband's attentions. A photograph taken of Carl, Emma and the children on one of the rare occasions the family were able to visit him during the war, when he was at the Château d'Oex in Vaud overseeing British POWs, catches this moment in Emma's life painfully well. There stands Jung, large, vigorous, the confident pater familias dressed English-style in plus-fours and tweed jacket, arm round his son Franz, now aged nine, with three of his daughters ranged alongside, each one as interesting and characterful as the next, looking direct and unsmiling at the camera, with Lil off camera, presumably in the care of the children's nurse as she was only two. And there sits Emma, quietly elegant in long dark skirts, a white broderie anglaise blouse and dark jacket, brooch at the neck, feet neatly crossed, head bowed, her face in shadow hidden by the rim of her wide straw hat, almost invisible, hating to be photographed – a picture of misery.

'Dear Gretli,' Carl wrote cheerfully to his second daughter, in the Basel dialect:

Many thanks for your sweet letter. I have a lot to do. Today I have to go back up to the Bernese Oberland, way high up to the glacier. There are lots of Englishmen there. Tell Agathli that we have only two Gurkha officers left. They are all brown and have heads wrapped round with a turban. Yesterday I was up on a high mountain and we got horribly rained on. I was with an Englishman who had been imprisoned in Germany for almost three years. While we were climbing the mountain we saw hundreds of black salamanders. They all sat there on the path and looked at us as if we had lost our wits to be out walking in weather where only salamanders are out.

Many kisses from Papa

Carl seemed to be having a fine time at Château d'Oex, living in a comfortable hotel, in charge of the kind of English he liked – officers and gentlemen – far from the horrors of war. The food was not bad, the company agreeable, conversation being in English, and there was even free time for reading and cycling and hiking in the Alps. But Personality No. 2 was never far away.

'While I was there I sketched every morning in a notebook a small circular drawing, a mandala, which seemed to correspond to my inner situation at the time. With the help of these drawings I could observe my psychic transformations from day to day,' he wrote later, recalling his inner struggle to understand what was happening to him at the time. Day after day he drew his mandalas, dozens of them. 'My mandalas were cryptograms concerning the state of the self which were presented to me anew each day. In them I saw the self – that is, my whole being – actively at work. To be sure, at first I could only dimly understand them; but they seemed to me highly significant, and I guarded them like precious pearls.'

Toni Wolff was a frequent visitor at the Château d'Oex. Of all the people surrounding Jung at that time only Toni was able to follow him into his symbol-laden unconscious. Emma tried, but found it impossible. She was less 'spiritual' than Toni, less intellectual perhaps. 'Toni was so important for him; she had an understanding for it,' said Toni's sister Susi. 'She perhaps was able to encourage him and so gave him the faith in himself. Frau Jung wasn't able to do it.' It was Toni he needed when, battling with his unconscious, teetering on the edge, he worked on 'Transformations of the Libido' and *Liber Novus*. 'Phenomenologically one might classify it as a schizophrenic episode,' thought Carl Meier. 'But it was voluntary. He simply had the guts and courage to deal with these contents which nobody has experienced except schizophrenics.' In Meier's opinion, this was Jung's real pioneering work. And living proof of Jung's theory that crisis need not be negative, might in fact be the very thing which led to development and healing: 'creative illness' as it was later described. And it was Toni Wolff who was his confidante then, not Emma. 'Going into the depths, no, that wasn't her kind of fish,' according to Meier.

Though still living at her mother's, Toni now seemed to be at Seestrasse all the time, helping Carl with his research, joining the family for meals and teas, followed by long private walks and talks with Carl in the garden during the summer months, long hours shut away with Carl in his *Cabinet* during the winter. Carl and Emma and Toni. Toni and Emma and Carl. The children hated it more than ever. Tante Toni appeared in their home as if by right, interfering in their private family life, taking their father away from them, making their mother unhappy. And their mother did nothing about it. It was beyond belief.

The only meal Emma managed to keep sacrosanct was the midday meal, when Carl would emerge from seeing his patients in his *Cabinet* and take his place at the head of the table in the *Stube* overlooking the lake. As usual, he was sometimes totally silent, lost in his own thoughts, when the children had to sit quiet and stop fidgeting; or at other times loud and exuberant, full of stories and jokes and tricks, making everyone laugh, causing havoc. He might tease the maid, flick peas and meat across the table with his fork, or shout at the cook if he felt the food wasn't hot enough, the meat too tough, the vegetables overcooked, leaving the cook in tears. It was like having a sixth child. '*Also, hor uf Carl!*' Emma would remonstrate: 'Stop it, Carl!' whilst the children fell about laughing. They soon knew when there was licence for bad behaviour and when not: if they had guests and they were getting bored they might squabble, or tell silly jokes or start to sing. Or if it was their favourite chocolate pudding for dessert, perform their 'party trick', taught to them by their father: pick up your bowl, hold it close to the lips, and blow the chocolate across the table, all over the crisp white tablecloth. Carl or Emma might admonish them for these misdemeanours but they were rarely punished, neither parent having the inclination to do much about it. The most that happened was a threat, rarely carried out, to be locked for a while in the small room under the cellar stairs.

But it was no laughing matter when their father got into one of his rages. These came out of nowhere like an Alpine storm, terrifying the children. Only their mother seemed unperturbed. 'He doesn't mean it. He can't help it,' she explained to Agathe when she was old enough to understand. It was true: the storms did pass, but it was hard for the children to understand. It just depended which Papa you had at any one time. It was the same in the evenings. If their father was in an introverted mood he might just go back into his *Cabinet*, or he might stay in the *Stube* – there but not there, reading one of his books, smoking his pipe, oblivious, whilst they sat at the table doing their homework or playing quietly at one of their board games. If on the other hand he was in an exuberant mood everyone was energised, up for some fun, whilst their mother, the only grown-up in the room, did her knitting or needlework, overseeing the chaos till bedtime.

Emma was always busy. First thing every morning, once the children set off for school after a rushed breakfast – 6.15 in summer, 7.15 in winter, Swiss style – she dealt with the servants, formally, as was the custom: the cook, the three maids, and the gardener-cum-handyman who brought in the daily vegetables and flowers and went shopping with his wooden cart into Kusnacht village for groceries or fresh fish and Herr Doktor's tobacco, but never the meat which came by post from the family butcher in Schaffhausen. Then it was a long morning of correspondence, mostly business for Carl, often interrupted by one of his patients – placating them if he was running late, arranging their next appointment; then checking up on the midday meal before the children came running back from school, wanting to tell their mother about their morning, followed by the meal itself which might run this way or that, depending on Carl's mood. During the hot summer months it was often taken outside in the garden under the rose bower, but it was always rushed, the maids hurrying in and out with the dishes, because the children had hardly an hour. Then off they went again, along the country road, leather satchels on backs, past the farm through the meadow with the cows and along the lake to the village school, the girls in their pinafores and pigtails, Franz in short trousers and leather braces, barefoot in the summer.

Once a week a seamstress came to the house who sewed the 'whites' – underclothing – and twice a year a tailor arrived from London to keep Carl attired in the English style, though naturally not during wartime. For Carl, it was either his Swiss peasant clothes – dungarees or shorts, or baggy old trousers, and green gardening apron – or those of an English gentleman: tweed or linen jackets, shirts and ties, corduroy trousers and woollen cardigans straight from St James's. 'How lucky you English are because you have these aristocrats and we in Switzerland don't know whether to wear yellow boots with top hats,' as Jung once wrote to his English friend Maurice Nicoll.

Emma's afternoons were more of the same, mostly revolving around Carl: the appointments diary, the social diary for visits, receptions, lunches, relations, family friends, unexpected guests from abroad – everything passed through her hands as she sat at her writing desk by the window in the *Stube*. And always Carl himself, bursting out of his *Cabinet* between appointments, wanting to know what she thought of his draft article, his chapter, his latest idea. 'Papa was always full of ideas and they talked and talked and we weren't allowed to go with them into the garden or library. We stayed downstairs,' remembered Agathe. 'She was not neglecting us, but her chief interest was Papa. She was there for Papa.' Their mother was very intelligent, Agathe thought, and should have gone to university. 'I think she was very good for Papa because she helped him, and was terribly interested.' Franz agreed: 'My father always kept her in play.' The world might not know it, but the family knew: their father couldn't manage without their mother. He was always demanding her attention: where was his favourite pipe, he'd put it down somewhere, and where was yesterday's newspaper, and had she answered so-and-so yet? And tell Toni to come straight up when she arrives.

At Seestrasse the *ménage à trois* became a permanent fixture. Susi Trüb saw how hard it was for Emma: 'She tried very hard to accept it and to understand it. But it was impossible.' When the Jungs were invited anywhere Carl arrived with both Emma and Toni. It did not do his reputation locally much good, though what did he care – these narrow-minded Swiss. The Jung family didn't go to church either, though churchgoing was part of the very fabric of Swiss life, everyone dressed in their Sunday best, the church bells pealing out across lake and Alps. Carl held strongly to his own private, symbolic religion, but gave a wide berth to the official Protestant Reformed Church which had caused his father so much anguish. There was no grace before mealtimes either, the only prayers being said by Emma at bedtime, when she came to say goodnight and the children put their hands together and shut their eyes, asking their guardian angel to protect them through the night. It all added to the rumours about the strange goings-on at Seestrasse, and the children found that some of their school friends were not allowed to visit the house. The walk to and from school could be hazardous too, especially for Franz, because the two Morell boys from the farm next door often lay in wait with sticks and stones. Aggi and Gret were safer because they were girls. Not that Gret cared. She was the rebel of the family and whenever there was discord amongst the children you could be sure Gret was behind it.

On the way home from school they would drop in on Groma Jung. Groma Jung was always there, ready with a drink of milk and a bun. In fact she never went anywhere except to visit Seestrasse. It was one of the best things about her, especially as the parents were always so busy. Tante Trudi was there too, but one hardly noticed her because she was so quiet, almost invisible really. If it was cold or raining Groma Jung would be inside by the tiled stove, doing some knitting or darning. If it was warm she would be out in the garden, tending her flowers and vegetables. She was always full of talk, telling stories from the Bible, or ghost stories on dark winter nights that were so real they were frightening. She had plenty of visitors, mainly her Preiswerk relations from Basel, but the only time she left her home was when she came to look after the children during one of Carl and Emma's holidays, when she and Trudi moved into one of the upstairs guest rooms. Apart from that there was Saturday bath night at Seestrasse, when the gardener fired up the big boiler in the cellar, or bathing in the lake in the summer, and the occasional Sunday lunch. Otherwise Groma Jung was happy not to budge.

Emma's central dilemma – how to manage life with her husband and how to find her own way through it – remained. Whenever she talked to Carl about it he always said the same thing: she had to shape her own life, develop her own way, stop relying so heavily on him and the children, analyse her dreams and her feelings, and pursue her studies. *Individuate*, as he termed it, summing up. It was Carl's suggestion that Emma start an analysis with Toni's brother-in-law Hans Trüb, Susi's husband, who had become one of Carl's students after training as a doctor. And so she did. The analysis covered many things: the problems in Emma's life and marriage certainly, but also their mutual interests in literature, mythology, religion, history. Emma and Hans found they had much in common, not least because they were both married to errant spouses. Susi Trüb was wild and sexually free, way beyond the norms of the times and in sharp contrast to her sister Toni. She was having an affair with a wealthy Russian émigré in Zürich named Emilii Medtner, who was one of the Jung circle. But he was by no means the only one.

Carl was often a guest for lunch or dinner at the Wolff house in Zürich, usually following a seminar or lecture, after which he would retire with Toni to her room. Whether it was a full sexual relationship remains a moot point, but the spiritual side of the relationship was what mattered most to Carl. Their mother knew about the relationship, said Susi, and it troubled her: 'But I think perhaps she was more troubled for Emma. Perhaps she thought it was good for Toni to have a "friend". But for Emma . . . She felt for Emma. All those difficulties!' Everyone in their circle knew about it. 'This was an exceptional relationship, because of the quality of the three persons involved,' thought Tina Keller, wife of Adolf Keller, pastor of St Peter's Church in Zürich. 'As a spectator I felt it was very different from an "affair" as I saw them around. There was responsibility and a common task which was beyond but included the love relationship.' The Kellers never invited the Jungs for supper without inviting Toni as well. When Tina arrived for an analytical session one day at Seestrasse, Jung said: 'This is how we three struggle with this problem.' He was at his desk in his *Cabinet* painting a manuscript illumination for his *Liber Novus* of three snakes intertwined.

Others in their circle saw Emma's distress: 'There isn't the slightest doubt in my mind that this relationship was a torture and a painful thing for Frau Jung to bear,' recalled one friend. 'He was so powerful.' When Carl and Emma entered a room everyone turned to Jung, drawn like a magnet, whilst Emma stood quietly in the background, saying little. One friend called Emma 'Mona Lisa' because of her quiet beauty and slow, knowing smile. Everyone liked and admired her for the dignified, intelligent and courageous way she handled herself. The only people who could have told you different were her maids. Every now and again Emma blew, and when she blew it was a flame of fury out of control and way beyond the misdemeanour which occasioned it.

The Jungs had a close-knit group of friends, built up over the years of their marriage. Many had children the same age and they all shared an interest in psychoanalysis. During the early years of the war, stranded as they were in neutral Switzerland, they spent many evenings together in each other's houses: the Trübs, the Kellers, the Pfisters, the Riklins, the Siggs (Martha Boeddinghaus, Carl's old admirer, had married a successful local businessman named Sigg), the Maeders, Maria Moltzer, Emilii Medtner, and of course Toni Wolff, with Seestrasse always at the heart of things. It was an interesting group: liberal, well educated, combative. Here women were included in the conversation as equals, not left behind in the drawing room whilst the gentlemen retreated to the *Herrenzimmer* to drink their wine and smoke their cigars. Tina Keller recalled that in the early days Frau Jung would sit quietly on the sidelines, but as time went on and she saw how forthright the women could be, as much as the men sometimes, she joined in more and more, gaining confidence and finding she had opinions as strong as anyone else.

Tina Keller first heard about psychoanalysis through her husband Adolf, who had already been a member of Bleuler's study evenings at the Burghölzli, discussing the new ideas emanating from Professor Freud in Vienna. Tina's parents had warned her about the 'dangerous' company Adolf kept: 'in 1912 Psycho-Analysis was [regarded as] a very controversial issue', but that only interested Tina more, and the pair were married in 1912. Tina's upbringing had been similar to Emma's, with the limited education routinely offered to girls, except for a stimulating period attending Cheltenham Ladies' College in England, where the headmistress encouraged her young ladies to aim high. The suffragists were grabbing the headlines, marching with banners for women's rights, a movement which the *Neue Zürcher Zeitung* described as 'ridiculous'. Tina came home fired up, but soon found there was nothing more for a woman in Switzerland to do – but perhaps become a nurse and then marry, which she duly did. But like Emma she married contrary to expectation, determined not to lead a life of stifling leisure. Adolf, almost twenty years her senior, came from modest peasant stock: a good, straightforward, open man in contrast to her much more complicated self. She was soon pregnant with their first child, and had four more children in quick succession, like Emma. After her third child she suddenly and unaccountably became profoundly depressed, filled with irrational fears and anxieties. Adolf worried that his wife had gone crazy and in 1915 brought her to see Jung for what would turn out to be a prolonged spell of analysis.

Jung told Frau Keller what he told so many of the intelligent, frustrated women who came to see him: she wasn't ill, she just needed to get away from her restrictions, have the courage to live her life more fully, more creatively, to individuate. Note all your dreams, he told her, and write down your thoughts in letters to me – not to send, just to express them fully. Jung was always good on practical advice like that, well ahead of his times in this as in other ways. She was to embrace her dark side too: 'God has a dark side,' he assured her. It was a good thing, not a bad. When Tina found the effects of the analysis were causing trouble in her marriage he told her it had little to do with the analysis; her depression showed she was already embarked on change.

It was during Frau Keller's visits to 228 Seestrasse for her sessions with Jung that she and Emma became friends. As the two women talked, they discovered they were both keen to further their education and make something more of themselves. Soon they had decided, with Carl's active encouragement, to employ the services of a professor to teach them maths, physics, and in Emma's case, more Latin and Greek. Emma was changing. Forced by difficult circumstances, trying to come to terms with the realities of her marriage, she finally began to take her life into her own hands.

13

The Americans

America and Americans had played an important part in Jung's working life ever since his first visit to Worcester with Freud and Ferenczi back in 1909. He had fallen in love with the place, the energy, the sheer newness and vastness of it all, and America in her turn had fallen for Dr Jung of Switzerland, the coming man, with his own brand of vitality and a bagful of challenging views, not all of them congenial. As a result there had been a stream of American physicians and patients crossing the Atlantic on the great ocean liners, making their way to Jung's door, first to the Burghölzli, then to Seestrasse. None spoke Swiss German but both Carl and Emma spoke English well enough, so conversation could still flow around the table over lunches and suppers or tea on the terrace, everyone throwing in their penny's worth, but dominated as ever by Jung.

One couple who became key figures in Carl and Emma's circle of friends was Edith and Harold McCormick. Edith was the daughter of John D. Rockefeller, the Standard Oil baron, reputedly the richest man in America. Edith was headstrong and outspoken, a keen reader with intellectual aspirations. In 1895 she married Harold McCormick, scion of the International Harvester family, thereby joining two of the wealthiest dynasties in the United States. Harold was a Princeton man, more sporty than intellectual, a clubbable genial sort, good-looking and a bit of a dandy. They lived in Chicago in a vast lakeshore mansion where they entertained lavishly, American style. On Chicago opera gala night Edith would arrive bedecked in the Rockefeller pearls, which the newspapers assured their readers were worth $2 million. Their opera dinner which followed made even splashier headlines than the pearls. They were the golden couple, but three years into their marriage they were hit by tragedy: their baby daughter died, aged one. A few years later their three-year-old son John died of scarlet fever, for which there was still no cure, all the money in the world unable to save him. Edith plunged into a depression with bouts of agoraphobia. There followed years of traipsing about the world to see the best medical specialists, but nothing helped – until they met Dr Jung of Switzerland on his trip to America in 1912.

The introduction came through Harold's cousin Medill McCormick, from the *Chicago Tribune* side of the family – the one who had been a patient of Jung's since 1908, suffering from alcoholism brought on, Jung suggested, by a domineering mother. Jung thought it had driven him to 'a wild and immoral life', as Medill wrote to his wife Ruth, and had warned him against being 'too good', recommending a little flirtation. He had just had one himself. But Jung also suggested that in a marriage 'love could suppress the former immoral tendencies'. So Medill recommended Jung to his cousin.

Edith had just finished another treatment, a fresh-air 'cure' in the Catskill Mountains. Jung found her lively and mentally alert but emotionally fragile. For her part, Edith was soon convinced that Jung was the man to save her and, accustomed to buying everything she desired, offered to set him up in America, including bringing his family over to live nearby. It must have come as a surprise that Jung refused her offer – not only refused, but insisted that if she wanted to be treated by him it was she, not he, who had to up sticks. Jung saw her again during his next trip to America in 1913, and this time she accompanied him back to Europe, together with her three surviving children, Muriel, Matilda and Fowler, and mountains of travelling trunks and hat boxes. Harold stayed behind in Chicago working in the family business. Edith set herself up in a large suite at the Baur-au-Lac, the grandest hotel in Zürich, ostensibly for a few months. She stayed for eight years.

By October 1913 it was clear to Harold that Edith was not coming back, so he too embarked for Zürich, to check on his wife's progress. 'Edith is becoming very *real* and *true to herself*,' he wrote to his father-in-law Rockefeller in December, 'and is seeking and I am sure will find her path . . . At any rate, she is in absolutely safe and trustworthy hands for no finer man ever breathed than Dr Jung. He has an intense admiration for Edith and yet recognises she is the toughest problem he ever had to deal with.' One of Jung's approaches was to encourage Edith to lead a more 'normal' life. Edith had a travel phobia and usually went everywhere in the safe confines of her chauffeur-driven automobile. Now she agreed to take the train from Zürich for her analytical sessions in Kusnacht – even buy her own ticket – but only if her chauffeur drove the car alongside the tracks so she could abandon the task if her panic grew too severe. Jung offered practical advice again: walk more, he told her, preferably hatless to abandon formality, and learn to knit (she never mastered it) and tidy your own room, which Mrs McCormick duly did, even helping the hotel chamber maid to clean the floor, which must have come as a shock to the maid.

Edith found 'normal' family life difficult. To simplify matters, Muriel and Matilda were each given twelve identical outfits, one set in pink the other in blue. By 1914 they were enrolled in local schools while Fowler plied back and forth to America, where he attended Groton prep school in Massachusetts. Years later Fowler still remembered Herr and Frau Doktor Jung coming to dine with his parents at the Hotel Baur-au-Lac and Jung teasing his wife when she, newly liberated, smoked a cigarette after the meal. Beaming, Jung invited everyone present: 'Now please look at Mrs Jung – how she is working with this new toy she has.'

Harold returned to Chicago and the family business but came back to Zürich in September 1914, by which time war had broken out. He kept the family in America up to date with Edith's progress in his chatty letters. 'Her step is springy and she walks with her arms free and swinging. She notices all things of nature and dresses simply and in very artistic taste,' he wrote to his mother. 'In the mornings we usually take a walk before lunch and in the afternoon also. Then in the evening we sit around the hotel or go to some moving picture show.' By October, Edith was deemed cured and able to turn her attention to other matters: astrology, biology, history and music – and training to be an analyst. Harold meanwhile entered analysis with Jung, as did Muriel, who had already spent some time in local sanatoriums. And the McCormicks joined Carl and Emma's circle of personal friends. They had a lot in common, especially with Emma, whose family had made their fortune in agricultural machinery just like Harold's. In August 1915

Harold joined Carl on a walking tour through the Swiss Alps, but only for a few days because Carl said he needed solitude and time to meditate.

Even the Jung children could see that Mrs McCormick was eccentric. But her hats were marvellous. As soon as she disappeared upstairs to see their father or into the *Stube* for one of their parents' receptions, they raced to the garderobe in the front hall and tried them on – huge hats decorated with all manner of feathers and veils. Papa's hats were there too: his flat straw hat, battered, his old grey Filtz, the black melon bowler hat, a sou'wester from America, a black silk top hat, and later, when he came back from Africa, a sola topi which he liked to wear on their Sunday afternoon walks. Mama had plenty of hats too, since women never went anywhere without a hat, but none as dramatic as Mrs McCormick's. Apparently Mrs McCormick was not the only eccentric in the family. When Papa came back from America he told them that her father had a bodyguard who followed him everywhere, because the tycoon had hurt so many people in business he had to be careful. All the gold in the world could not stop Mr Rockefeller being a lonely, suspicious old man, said Papa.

It soon came to the McCormicks' attention that Jung needed a proper place to give his talks and seminars, which usually took place, somewhat impromptu, in the private rooms of the Seidenhof restaurant in Zürich. They suggested establishing a club, which Edith was more than happy to fund. A substantial property was found in the centre of town at 1 Löwenstrasse, and by early 1916 it was set up. 'I am enclosing a photo of the Psychological Club which I founded on 26th of January this year,' Edith wrote to her mother-in-law. 'The house I have rented for two and a half years. It makes a centre for analysed people where they can be in *pension*, or come in for meals, or come for the evenings for lectures, discussions and study, all of which teaches collectively. Any new movement has a slow growth, but this assures a lasting quality.' To fund it further she asked her father to increase her allowance, currently at $2,500 a month. Whatever he gave her was not enough, apparently, as she subsequently took out a bank loan of $80,000. 'Pfister writes that Rockefeller's daughter presented Jung with a gift of $360,000 for the construction of a casino, analytic institute etc.,' Freud wrote sourly to Sandor Ferenczi on 29 April 1916: 'So Swiss ethics have finally made their sought-after contact with American money. I think not without bitterness about the pitiful situations of the members of our Association.' No wonder he was bitter. Times were hard, not only for Viennese psychoanalysts. Austria was at war. All three of Freud's sons were enlisted, fighting for their country.

Despite the lavish funding, there was trouble at the Psychological Club from the start. First it was Alphonse Maeder, a colleague of Jung's since early days at the Burghölzli who had become the first president of the Swiss Gesellschaft für Psychoanalyse when the Zürich group split from the Viennese in 1913, but who now refused the post of president of the new club, protesting he would be no more than a mouthpiece for Jung. 'He was too strong and dominant,' Maeder said later. 'He was a man of great stature, surely a genius personality, but strong and a bit massively Swiss German.' Maeder knew he couldn't stand up to him. Meier noted: 'Jung was very critical of men. It was frightening almost. There was something odd about it.' Maeder agreed: 'He was not able to maintain a real friendship among men. He had some, but rather weak ones. Those were men who admired and followed him. He had intelligent women, crowds of women, around him, but with men he could not really get along.'

It was true: there were few men of real stature in Jung's group, and he was not above bullying them either when the mood took him. Tina Keller came to hate the way Jung could behave in company, magnetically charismatic one minute, 'vulgar and repellent' the next, making fun of someone sarcastically and unfeelingly, picking on the weaker ones, and swearing as crudely as a Swiss peasant. In that mood split Carl could frighten people and Emma often had to step in to console and conciliate. Maria Schmid, the wife of Jung's friend and colleague Hans, with whom he often went bicycling, remembered Emma Jung saying she was rather sorry that her husband had no real friends; Hans was one of the few people Jung addressed as *Du* rather than the more normal and formal *Sie*. Alphonse Maeder decided to protect himself and ducked out. But who should take on the job instead? The answer was Emma.

The timing could not have been better: Lil was nearly three, the other four children were out at school most of the day. Emma was just beginning to develop her own life and here was the perfect opportunity. Encouraged by Carl, she decided to take it on. It was something of a turning point, even though most members thought, like Maeder, she would be little more than a mouthpiece for her husband. The first meeting was held on 26 February 1916 with forty people present, twenty-four of whom were women. Emma was approved as president, their friend Hermann Sigg, with his business experience, was appointed treasurer, and Irma Oczeret, a Hungarian who was part of their analytic circle, secretary. The second meeting was held on 15 March, when fifteen further members joined. Almost immediately, however, there was more trouble.

It lay in the fact that the club did not really know what it was. The McCormicks saw it as a club along American lines, social as much as a place of study, with rooms for *pension* on the upper floors, and on the lower floors a dining room, billiard room, games room including a table-tennis table, library, lecture room and big reception room, with a variety of smaller withdrawing rooms where analysts could see their patients. The Swiss on the other hand saw it essentially as a serious meeting place for talks, lectures, analytic sessions, and relevant social functions; a place for an informal exchange of views, certainly, but hardly a place to attend in order to play ping-pong. The club was serviced by a staff of three servants, a cook, maintenance men and a concierge, all overseen by a Fräulein Teuscher. Toni Wolff was one of the first to point out that it was too luxurious for their local members, the restaurant too expensive, and likewise the accommodation. As a result it was rarely used during the week unless there was a talk by a club member or one of Jung's sell-out lectures or seminars. By July 1916 the club was in financial trouble, only solved by Edith McCormick pressing her father for more funds. There was also the problem of rank: who was a full member, who an associate. Analysts formed one group, analysands the other. Harold McCormick, the sociable club-bable man who once entertained the assembled with a whistling serenade, noted that 'unconsciously there is too much atmos-phere of rank in the Club . . . the mantle of "caste" should be laid aside at the threshold of the Club and the Natural Simple Human Relation assumed in its real aspect'. His private view was that the Swiss did not know how to have some plain good fun. But some of the plain good fun – the jolly Christmas party and the fancy dress balls, some of the evenings of games and the summer excursions to the Swiss Alps and lakes – was open exclusively to full members, leaving associate members out in the cold.

Jokey Harold McCormick with his daughters in Zürich during the First World War.

Jung himself was clear about the essential purpose of the Psychological Club: to overcome the limitations imposed by one-to-one analysis, another forward-looking view anticipating later developments in group therapy. He thought both aspects of club life, intellectual and social, could contribute to this – in fact it rather neatly fed into his new concept of introvert and extrovert, which he liked to try out on members by way of research for his book on psychological types. But the rows and disagreements would not go away, bedevilled as they were by jealousies and rivalries, everyone competing for Jung's attention. It was the first test of Emma's leadership, and in October 1916 she issued a circular letter about 'the Club problem' to all members. In her role as conciliator she asked all 'who have a real interest in the development of our Club' for their 'impressions, encouragement and criticisms both in matters of principle and practical'.

If anyone thought the letter had been dictated by Jung they only had to look at the language to see that it was pure Emma. 'If the Club hasn't as yet become as great a boon as we hoped, this doesn't mean it won't become one in the future. In my view it takes time to establish a thing . . .' Emma was liked and respected by everyone. Most members responded to her appeal, though not everyone in the same spirit of optimism. 'My suggestion would be: an absolute reorganisation of the Club,' wrote Maria Moltzer, now a practising analyst, but disconsolate that she had been given no formal role to play, 'since it seems to me that the present club is incurably ill.' It was her view that the organisation, financially supported by Edith McCormick, was parasitic. Fanny Bowditch, an American who came to Zürich to be analysed by Jung after she had fallen into a deep depression following the death of her father, wrote to 'My dear Mrs Jung' saying she was interested in the club's welfare but she found herself unable to take an active part because she needed solitude more than collectivity. Used to having Jung to herself in their private sessions, and half in love with him, she later admitted to Herr Doktor that seeing him in a group was not to her liking: 'my eyes were opened to the reality of things and I saw you in a new light – for the first time, in the grip of your own complexes'. She was not the only one. No wonder Jung wanted a club 'to overcome the limitations of one-to-one analysis'. No wonder Emma did too. Jung soon passed Fanny Bowditch over to Maria Moltzer. Emma's efforts prevailed: within three years the Psychological Club had moved to a more modest house in Gemeindestrasse, where it still functions to this day.

When it was Emma's turn to give a lecture at the club it required courage to overcome her natural shyness. As her subject she chose 'Guilt'. Even in the English language this is a challenging word, but in German '*Schuld*' has several meanings: guilt, blame/fault, debt. Everyone in the room knew about the *ménage à trois*. Once, when Alphonse Maeder slighted Toni Wolff at the club, Jung stormed out of the room, furious, red-faced and shouting: 'I'll take her on my knee and hold her there throughout every meeting till they stop hounding her.' So now here was Emma standing at the lectern, facing the members, speaking about guilt, with Carl and Toni in the front row in the armchairs always reserved for them.

She began with etymology, an interest she had pursued from the early years when she had first helped Carl with his work: in French there's '*devoir*' and '*conscience*', in English there's 'should', in German '*Pflicht*', duty. Already in mythology the word 'guilt' is linked to suffering, she said. But without suffering there could be no development, no growth. 'I myself have had to grapple with this,' she added, surprising everyone in the room with her boldness. The feeling associated with guilt was often fear, she went on, bolder and bolder. But the main interest in the word 'guilt' was psychological, which in turn depended on the context, the time and the place. She offered some history, starting with the cleansing ceremonies in primitive cultures, warding off evil. The feeling of guilt could come when a person was not 'at one with himself'. Was she looking at her husband when she said it? By his own admission Jung suffered from a 'constant feeling of guilt'. On Emma went, addressing the Christian concept of 'sin', a word which cropped up again and again in Jung's writings: the forbidden fruit of Adam and Eve; Pandora opening her dreaded box. She ended with the central message of the lecture: we have to take responsibility for the sufferings in our own lives and solve our dilemmas for ourselves. You can almost hear the clapping.

Emma's friendship with Tina Keller was put to the test during this time when, in a classic case of transference, Tina fell passionately in love with Carl during her analysis. But Emma stayed steady and Jung handled the situation with 'respect, tact and sincerity', telling Tina: 'You must be quite clear that you are alone in this experience.' Always believing that analysis was an equal give-and-take between the analyst and the analysand, Jung admitted to her how he sometimes felt close to insanity. And Tina could see for herself how deeply he was still affected by his split from Freud because he could never talk about Freud without his face suffusing with emotion. 'My father would not confess it,' his son Franz later reflected, 'but he probably never got over Freud in all those years.' Alphonse Maeder thought it went back to Jung's own father. 'Jung [was] religiously interested. But religious in a somewhat other sense than Christian or ecclesiastical religion . . . Even up to his last day he had a complex against the Church and her mission. He would never use the word "Church" without swearing: it remained a real father complex.'

Emma knew her husband teetered on insanity at times and that he needed Toni Wolff to help him through his struggles. She was what the Jung circle of analysts came to term an 'anima' figure, a sort of muse to the unconscious. 'She had an extraordinary genius for accompanying men – and sometimes women too, in a different way – whose destiny it was to enter the unconscious,' recalled Barbara Hannah, the Englishwoman and member of the Jung circle who later wrote a biography of him. 'Toni Wolff was perhaps – of all the "anima" types I have known – the most suited to carry the projection of this figure.' Hannah saw how hard this was for Emma. 'What saved the situation was that there was no lack of love in any of the three. Jung was able to give both his wife and Toni a most satisfactory amount, and *both* women *really* loved him. Therefore, although for a long time they were most painfully jealous of each other, love always won out in the end and prevented any destructive action on either side.' Emma even said years later: 'You see, he never took anything from me to give to Toni, but the more he gave her, the more he seemed able to give to me.' Hannah doubted Jung could have survived this time of inner crisis alone.

Jung himself put it more breezily: 'It is unfortunately true that when you are a wife and mother you can hardly be the *hetaira* too, just as it is the secret suffering of the *hetaira* that she is not a mother.' The *hetairai* were the courtesans of ancient Greece, respected for their learning and beauty, and afforded greater freedom than married women. 'There are women who are not meant to bear physical children,' Jung went on, 'but they are those that give rebirth to a man in a spiritual sense, which is a highly important function.' Susi Trüb agreed: 'She wasn't at all practical,' she said of her sister Toni. 'She would never have been able to lead a married life.'

* * *

Jung's Personality No. 2 often caused havoc at at 228 Seestrasse. One day in 1916 everything came to a head. 'It began with restlessness, but I did not know what it meant or what "they" wanted of me. There was an ominous atmosphere all around me. I had the strange feeling that the air was filled with ghostly entities. Then it was as if my house began to be haunted.' Aggi saw a white figure passing through the room. Gret said her feather bedcover was snatched in the night. Franz had a nightmare. In the morning when Emma gave him some crayons he drew a picture of his dream: a fisherman on the banks of a river with a chimney on his head belching flames and smoke, landing a fish. From the other bank the devil came flying through the air, cursing because the fish had been stolen. An angel hovered above the fisherman, saying: 'You cannot do anything to him; he only catches the bad fish.' Franz called it: 'The Picture of the Fisherman'. Perhaps he knew about the story of the Fisher King his father had illustrated in the *Liber Novus*.

The next day at 5 p.m. the doorbell at 228 Seestrasse began clanging frantically. It was a bright summer's day and the two maids were in the kitchen which overlooked the front entrance. There was no one there. Everyone raced to look, but there was nothing. 'The atmosphere was thick, believe me! Then I knew that something had to happen. The whole house was filled as if there were a crowd present, crammed full of spirits,' wrote Jung. 'They were packed deep right up to the door, and the air was so thick it was scarcely possible to breathe. As for myself, I was all aquiver with the question: "For God's sake, what in the world is this?" Then they cried out in chorus: "We have come back from Jerusalem where we found not what we sought."'

The engagement. A few out-of-focus snapshots taken at Ölberg are the only record of Emma and Carl's secret engagement in October 1901. *Verlobung*, 'engagement', is written faintly on the reverse.

Emma on her wedding day, 14 February 1903, St Valentine's Day. There was a ball and a grand banquet, but Emma's father was too ill to attend either of these, or even the church ceremony.

Zürich c. 1905. Once married, Emma joined Carl in Zürich. While he worked a twelve-hour day, Emma ventured into the city, much more modern and busy than sleepy Schaffhausen.

Carl and Emma with Aggie and Gretli. Emma took the photograph of Carl, and Carl took that of Emma. These early years were happy ones for them both. Married for four years, Emma was already the mother of two. Within a year she would be pregnant again, with Franz.

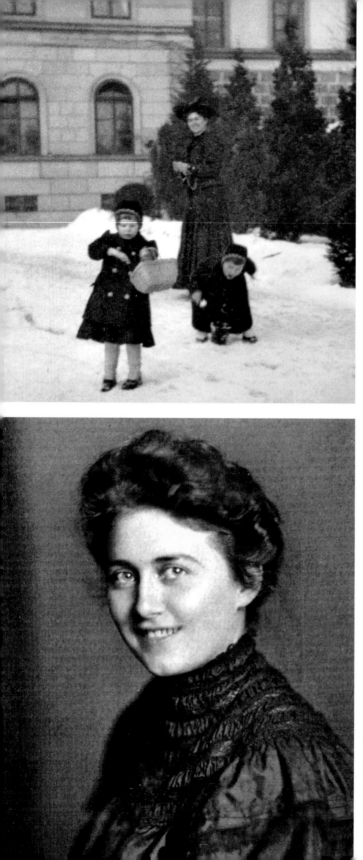

Emma with Aggie and Gretli outside the Burghölzli. By now she was taking an active part in the life of the asylum. Bleuler thought it a good idea to involve the wives.

A rare studio portrait of Emma, who hated having her photograph taken, preferring to stay in the background. It was Carl who enjoyed the limelight.

The Weimar Congress, 1911. All the leading lights of the early psychoanalytic movement were present, with Freud standing on a stool in the middle row, dominant. In the front row are Maria Moltzer (*third from left*) and Toni Wolff (*third from right*).

Emma (*front row, centre*) was encouraged to attend by Carl, who bends solicitously over her. More confident now, she was beginning to engage in the exciting new world of psychoanalysis.

Seestrasse 228, Küsnacht.
Emma and Carl finally
built their own home,
on Lake Zürich, in 1909,
when these photographs
were taken. Carl oversaw
every last detail of the
building, but Emma quietly
got her way too, and the
final result is a pleasing
compromise between
husband and wife.

Toni Wolff. Originally a patient of Carl's, she soon became indispensable to him. She had 'seer's eyes', and understood Carl's 'other' personality – the hidden, complex one.

Carl and Toni in England, 1922. Soon they were a *ménage à trois*, but Carl did it openly, flaunting bourgeois conventions. Emma was left to come to terms with it.

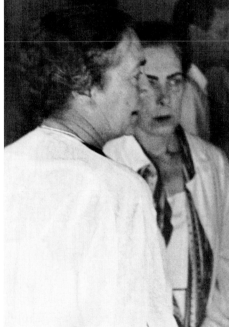

Emma (*left*) and Toni (*right*) at the Psychological Club, Zürich. By the 1930s Emma was practising as an analyst in her own right, and Toni Wolff no longer caused her the same anguish.

The Jung family in 1917 – only baby Hélène is missing. At this point in their marriage Emma was still having a difficult time.

Emma and Carl in the 1930s. By now they had come through their turbulent times to emerge triumphant. They could always laugh together, said their grandchildren, and 'both learned from each other all their lives'.

Jung went to his *Cabinet* and began writing *Septem Sermones ad Mortuos* (*Seven Sermons to the Dead*). 'It began to flow out of me, and in the course of three evenings the thing was written.' As soon as he started to write the whole ghostly assemblage evaporated, the room quieted and the atmosphere cleared. 'As day fades the unconscious is activated, and by midnight the censer is full blaze, but lighting up the past,' Jung said in one of his 1925 talks to the Psychological Club. 'As the dynamic principle increases in power, and the further back we go, the more we are overcome by the unconscious. Lunatics go further back to a strange psychological state where they cannot understand their ideas, nor are they able to make them understood by others.'

In a sense *Septem Sermones* is the distillation of everything Jung had been battling over the years: the spirits had not found what they sought. Now it had to be found: the integration of the personality, both good and evil, including the 'daemon of sexuality'. No more splitting. In the seventh and last sermon there is a star, shedding light. Later Jung conceded: 'No doubt it was connected with the state of emotion I was in at the time, and which was favourable to para-psychological phenomena. It was an unconscious constellation whose peculiar atmosphere I recognised as the *numen* of an archetype.' It was the same territory as his 1902 dissertation 'On the Psychology and Pathology of So-Called Occult Phenomena'. 'The intellect, of course, would like to arrogate to itself some scientific, physical knowledge of the affair, or, preferably, to write the whole thing off as a violation of the rules,' he wrote, back to his confident self. 'But what a dreary world it would be if the rules were not violated sometimes!' Are you listening, Freud? he might as well have written. He showed the text to only a few people, including Emma and Toni Wolff and Edith McCormick, and he did not publish it till a year before his death.

The place the Jung family spent most of their free time during those war years was at Ölberg, with Emma's mother, Bertha. Many weekends and Easters and the long summer holidays were whiled away there, and even if it had not been wartime and they'd been free to travel elsewhere it would still have been their favourite place. 'All summer long we played Indians against the English with my cousins. Papa was the leader,' remembered Franz. 'He wore a Canadian Mountie's hat and a pair of cowboy boots from his visit to America with Freud. He looked like a sheriff. We built teepees and huts big enough to sleep in, and each side had a horse. We would light fires and burn down each other's teepees and steal the horses. This was Papa's idea. He played with us all the time, although his brother-in-law did not approve.' Ernst Homberger, married to Emma's sister Marguerite, the hard, ambitious man running the family business, found Carl a complete anathema, hardly a man at all really: playing like a child, and always talking about something called 'the unconscious'.

The Hombergers had five children and the cousins, Homberger and Henne alike, joined in the wild games. Mostly Emma let them be – '*heb die Daumen*', fingers crossed that no one got badly hurt. Bertha Rauschenbach never worried: she loved adventure and thought it was good for the children. Every year she booked the ferryman to row the cousins, screaming with excitement and fear, to the rock in the middle of the hazardous Rhine Falls, or took them on steep single-track mountain railways high into the Alps, or more sedately on carriage drives with faithful Braun the coachman, to places of local interest. Marie always came too. Her father had been Jean Rauschenbach's carer during the last years of his life, and Marie herself had started as a housemaid aged seventeen. Now, thirty years later, she was Bertha Rauschenbach's faithful companion. They never quarrelled and they agreed on everything. But even they were alarmed when Carl got the cousins digging tunnels in the sandy soil, crawling through them by way of a daring escape from the enemy.

Summer weekends were often spent by the lake at Seestrasse. Everything centred on it: swimming, sailing the *Pelican*, Carl's yawl with red sails; rowing, digging dams and channels down to the lake from the stream by the vegetable patch, or making fires on the lakeshore to roast *Cervelat* sausages. And there it was always Carl who played with the children, never Emma. She preferred to sit on the terrace reading or doing her needlework, keeping an eye on things. If she was reading her *Nebelspalter*, the satirical weekly magazine, it could make her laugh out loud. Occasionally she took a swim, though rarely further than their own small harbour walls, whereas Marguerite swam far out. A wondrous event occurred on Saturdays when Groma Jung and Tante Trudi would arrive to bathe in the lake, swathed in voluminous bathing dresses, bearing bars of soap and flannels, because during the hot summers they preferred to wash there, not in the bathroom tub. And once every summer an old *Tante* arrived on a visit, one of Emma's many distant Schaffhausen relatives, and Carl would often take her back to Kusnacht by sailboat to catch the local train to Zürich and thence back to

Schaffhausen. One year, as the aunt stood balanced between jetty and boat, Carl offered a helping hand but she lost her footing and in she went, hat and all. Judging by his bellowing laughter it was one of Carl's little 'tricks'. The children howled with delight. Even Emma found it hard not to laugh. Luckily the old aunt was happy to share the joke.

In the evenings the family, like many Swiss, played *Jass* cards, and if not cards, then mahjong. They sat round the big table and played, fiendishly and noisily, till bedtime. The children knew from experience that when Papa was losing he was not above cheating. It was the same with badminton, where he unashamedly toed his opponent's shuttlecock outside the line when necessary. He always had to win. Once the children were in bed and if Carl wasn't working, he and Emma might settle down to one of their marathon games of patience, which ran not just for one evening but over many, pitting themselves furiously against one another. Or billiards in the veranda room which could, likewise, run for days. And somehow Carl always won in the end. Because, cheating or not, Emma always let him.

Christmases were spent at Ölberg or at home. Either way it was in the deepest snow, the garden, trees, lake, meadows, Alps, all white. At Seestrasse the Christmas tree stood in the far corner by the big fireplace, high as the *Stube* ceiling, lit only by candles and hung with stars and hearts baked by the cook. The crib with baby Jesus, Mary and Joseph, the three kings and the shepherds, stood on a table between the fireplace and the alcove with the two stained-glass windows commemorating Emma and Carl's wedding and the birth of Aggi and Gret. On Christmas Eve everyone dressed for dinner, Carl in black tie, Emma in evening gown and jewels, with the children, Emilie Jung and Trudi and sometimes Bertha Rauschenbach, all grouped round the tree singing 'Stille Nacht, Heilige Nacht'.

Groma Rauschenbach sitting at her usual table at Ölberg.

Then the servants, standing in a row at the back, were given their presents before returning to their work. The Christmas feast was followed by rowdy games such as musical chairs, led, as ever, by Carl. The next day the gardener Herr Müller's children would arrive to present themselves in the front hall, standing in a row from the oldest to the youngest, waiting for Frau Doktor Jung to come down the stairs and present them, formally, with their gifts. Then the children would curtsey and bow: '*Danke Frau Doktor, danke Frau Doktor*', before walking back home along the country road to Kusnacht village.

As the war entered its fourth year even safe little Switzerland was feeling the deprivations. Coal was so scarce the Jung house could no longer be heated by the boiler in the cellar. With his usual enthusiasm for camping and the outdoors, Carl fixed up a tent-like construction round the big marble fireplace and everyone crowded in there for warmth. As everything with Carl, it became an exciting thing to do. Petrol had run out long ago, for everyone but the McCormicks who did a deal with the American armed forces in Europe. But the Jungs were hardly struggling: the hens still laid their eggs, the lake still had its fish, Bertha Rauschenbach still sent the occasional joint of meat in the post from her butcher in Schaffhausen, and there was never a lack of vegetables because the lawn to the left of the drive with the swings and sandpits had been dug over to augment the vegetable patch, supplying the cook with all the fresh produce she required. Carl was in charge of the potatoes. Wearing his old gardener's overalls, he planted them, dug them up, kicked the earth off them and washed them in the stream before dumping them on the kitchen table like the good Swiss peasant he was at heart.

It was a different story for the Zürich poor. By 1917 they were close to starving. In November there were the first signs of unrest: *Hausfraus* in slippers and aprons queued outside shops complaining about the exorbitant rise in prices: bread up 19 rappen the kilo, milk up 6 rappen, flour up 20, sugar doubled in price, even potatoes cost 15 rappen a kilo more than in 1915. Forget coal, meat and wood. And all the while their husbands and sons stood at the Swiss borders doing their guard duty for 2 francs a day, with no assurance that their factory jobs would still be waiting for them when they returned. But up the Bahnhofstrasse, the cafés and restaurants remained busy, full of the idle rich, the ladies dripping with jewels. Life went on as usual in the Kronenhalle and the Odeon with those crazy expatriates. The rich of Europe and America still lived it up in the Hotel Baur-au-Lac, war profiteers among them. It wan't right. The political activists reminded them that Switzerland was still producing armaments for anyone who would pay. Down with war! Down with Capitalism! Workers of the World Unite! Look at Russia – do as they do!

On 17 November 1917 there was a demonstration in the Helvetia Platz. Barricades went up on the Badenerstrasse. The city police and then the military arrived. Shots were fired, resulting in four dead and seventeen wounded, scattering the protestors. And nothing changed. The following year, in September and with the end of war in sight, there were more strikes. In November there was a general strike. The country was hit by Spanish influenza, from the soldiers in the trenches they said. All schools, theatres, churches and public gatherings were shut down, but after more than four years of war the population had little resistance: 200,000 stricken in Switzerland, 900 died. The City Fathers, terrified by the spectre of revolution in Russia and now in Germany too, accepted that reforms were necessary. A forty-eight-hour week was agreed before the end of the year. But the following spring, once the immediate threat was over, the strike leaders were arrested.

Most of Zürich could not wait to get back to normal. Fashion was always of interest. 'The longer the war, the shorter the skirts,' quipped the *Neue Zürcher Zeitung*. Gerhart Hauptmann arrived to give a reading of his *Versunkene Glocke* in the Tonhalle; Herr Professor Einstein joined the University of Zürich; women's rights and the vote were discussed in the chamber. 'We know bayonets are not the solution,' said *Wissen und Leben* magazine. 'Zürich, the largest Swiss city, stands in a crisis of European origin. We stand on the edge of an era of social fairness, on the edge of a new Switzerland. Zürich is suffering, but has faith. And faith will triumph.'

At Seestrasse they had their own renewed faith, out of times of war into times of peace. Carl had spent 117 days on active military service in 1917, but now it was over. And so, more or less, was his inner crisis. As was one of his long-running infatuations. 'It was only towards the end of the First World War that I gradually began to emerge from the darkness,' he wrote. 'Two events contributed to this. The first was that I broke with the woman who was determined to convince me that my fantasies had artistic value; the second and principal event was that I began to understand mandala drawings.' Emerging from his darkness, he saw their true significance: the way they expressed 'wholeness'. As to 'the woman' he broke with: that was Maria Moltzer. The two events went hand in hand. Because Maria Moltzer failed to understand that his paintings were not aesthetic but symbolic – a route to the unconscious. That is, she failed to understand the core of his being. It was the end of things between them. Soon she had resigned from the Psychological Club and was effectively gone from Carl and Emma's life.

14

Into the Twenties

'London, 1 July 1919. Dear Marianne. It was sweet of you to write me a letter. It has made me so happy that I am writing a letter to you too. If you can't read it, Mama will read it to you. I have bought a doll here. It is carved from brown wood and comes from India. But it is for Mama,' wrote Carl to his eight-year-old daughter. In Switzerland children did not go to school till they were seven, so perhaps Marianne's reading was not yet good enough, but Carl's handwriting was more likely the problem.

As soon as possible, once the war was over, Jung had escaped the enclosed world of Zürich and made his way to London, where he had been invited to lecture to the Royal Society of Medicine and the Society for Psychical Research. He was staying near the King's 'Castle', he told Marianne, and there was a Tower where the Crown Jewels were kept. 'London lies on a big river where the seaships go,' he continued. 'Every day the river flows downwards for 6 hours and then upwards for 6 hours. Just think, more than twice as many people live in London as in the whole of Switzerland. Chinamen live here too. Many loving greetings to you and Lilli, from your Papa.'

Marianne was the musical one in the family: like her mother, she played the piano and the guitar too, and she had a fine voice.

When she left school she did a course of bookbinding in Paris and took some music and language courses in London. Later she was the one to show an interest in her father's work and acted as his secretary for some years. Aggi hated school and couldn't wait to leave. She resented her mother's decision to send her to the 'horrible' *Mädchenschule* in Zürich, the private young ladies' school, rather than the *Gymnasium* which was co-educational, with Emma concerned that her daughter was more interested in boys than her studies. Considering the unconventional way their parents chose to lead their lives, it is striking that the way they brought up their daughters was so conventional. All four daughters were sent to French-speaking Switzerland for a year when they left school, to improve their French and learn something of the 'art of housekeeping', because their parents thought their daughters' futures lay in being good wives and mothers. Further education could come later, if at all, as it had with Emma.

In 1919, aged fifteen, Aggi had already met the man she would marry, Kurt Niehus, a friend of one of the Preiswerk cousins and an engineering student at Zürich's famous ETH, the Federal Institute of Technology. They married when she was nineteen. Meanwhile she settled down to life at home because young women of her background rarely went out to work. Instead she helped with the younger children, attended some courses at the university, and spent time with her grandmothers in Kusnacht and Schaffhausen, filling in time until she could marry Kurt and lead her own life. Marianne and Lil were thrilled. Aggi was sweet-natured and happy to play with them, giving them rides on the back of her bicycle, swimming with them in the lake, taking them for walks, or if it was raining, sitting round the large table in the *Stube* pressing flowers or drawing or playing with their set of wooden animals, whilst Franz raced around playing with his train set. In the evenings, if her parents were engaged in one of their marathon card games, Aggi might help the children's maid get the younger ones ready for bed, brushing their hair and reading them bedtime stories.

Gret's memory of their parents was different to Aggi's because she was closer to their mother and never really got on with their father. It annoyed her that her mother always defended him, even when he got into one of his rages, saying he didn't mean it, he couldn't help it, and it would soon pass. Once the girls were married and had homes of their own it was always Aggi their father liked to see, rarely Gret. She irritated him too much. Ever since she was small they had rubbed each other up the wrong way because Gretli was strong and wilful and moody, much like Carl himself. She was interested in astrology and said she wanted to go to university to study psychology, but she wasn't allowed. Perhaps she should have been the boy. Franz was the 'soft' one, the one who took after his mother's side of the family, as Carl often pointed out. Being the only son, Franz found himself in a difficult position: on the one hand his father spent much more time with him than the girls, taking him on Alpine hikes, or sailing, or even letting him sit and draw beside him whilst he worked on his *Liber Novus*. On the other, he felt awed and dominated by his father and had to find ways of holding his own. By the 1920s Carl had acquired a collection of boats: the *Pelican*, two smaller sailing boats, as well as a rowing boat and a canoe. Franz refused point blank to crew for his father in the yawl, preferring to take the helm of one of the smaller boats. It was no fun sailing with Papa, who shouted orders at everyone and raged if someone did something wrong – until they were out into the lake at which point he relaxed, filled his pipe, and started telling his magical stories. But as far as Franz was concerned it wasn't worth it. He left crewing for his father to his Preiswerk cousins and the Niehus boys, Kurt and his brother Walter, who later married Marianne.

Perhaps Lil was the lucky one. Youngest of the five, she had what she remembered as a happy childhood in the safety of their home in Seestrasse, surrounded by the big garden, playing with the dogs – because Papa always had dogs – swimming in the lake, and going for visits to Groma Jung or Groma Rauschenbach in Schaffhausen. Lil spent a lot of time with Müller, helping him in the garden, collecting the eggs from the henhouse or accompanying him on shopping trips to Kusnacht village, pulling the little wooden cart behind them. Müller had been with the Jung family for ever, having started as 'young Müller' and ending as 'old Müller' forty years later. Looking back, Lil recalled that though their mother never kissed them, except on birthdays or at Christmas and Easter (few Swiss mothers did in those days), Emma was a warm-hearted and thoughtful mother. But busy, always busy, with the large house to run and countless visitors to entertain, including those coming from England and America. Famous ones too sometimes, such as Albert Einstein, Hermann Hesse and James Joyce, who came to see Jung about his daughter Lucia's troubles. But Emma still found time to listen to her children's problems and questions, sitting at her desk in a corner of the *Stube*, which was her base, the heart of the house. Any subject was allowed. 'She had a natural authority,' remembered Lil. And a good sense of humour too, laughing easily, especially when Papa told one of his outlandish stories. But when she was discussing work with their father she could be quite critical, observed Lil, which other people never dared. So her mother's occasional uncertainty came as a surprise. It was the same old uncertainty Emma had to deal with all her life: with strangers she was reserved and quiet, appearing a bit grand, 'always the lady', preferring to stay in the background. So many people got the wrong impression of Frau Doktor Jung, seeing her as little more than the wife and mother. But none who knew her well.

Aggi and Lil both remembered animated mealtimes, lively family festivities, and lots of card-playing in the evenings. If their father was in one of his good moods, full of vitality, transforming everything, then the fun would begin. 'Father was often terribly funny. We used to talk silly things and laugh like anything – terribly silly things sometimes,' Aggi said about the midday meal. Their mother didn't like it, but the children would repeat the silly words again and again, 'until we finally nearly rolled on the floor with laughter. Father loved such things. I can still remember mother saying: "It's really too stupid."' She did not like it either when Carl started feeding one of the dogs under the table, teasing them and working them up into a state. Even worse were his eating habits, because Carl had not been brought up with the same manners as Emma, and he had no intention of changing either. Swiss peasant was his style. 'The way he used to sip or eat his soup or soft boiled eggs was simply disgusting, one of the most disgusting things you could see in your life,' one of Jung's grandchildren later recalled. 'He made such horrible noises when he ate. Many times my appetite was completely spoiled. Of course nobody dared to say a word of criticism. When you saw him and my grandmother eating – she was such a gracious and refined lady – it was the most grotesque contrast.

He was aware of this and he just used to say "one has to eat one's food the way one enjoys it most".'

Jung was not home for long after his 1919 trip to England before he was off again. Over the next five years he went to England three times, America twice, Germany, France and Holland, and now, in the spring of 1920, he went to North Africa, invited by his friend Hermann Sigg, who had to go there on business, the business being oil. Jung called it his *'syndrome ambulatoire'*, explaining that 'in such cases people are seized with something like an urge to travel, have an amnesia concerning the past and just travel or run away'. Partly it was wanting to get away from the pressure of work. There were now so many patients arriving from England and America to add to his local ones that he often started work at 7 a.m. The patients stayed at the Hotel Sonne in Kusnacht village, sitting on the terrace overlooking the lake every morning, writing out their dreams of the night before. 'Jung unfortunately had a great success in his London lecture,' Ernest Jones had written to Freud in the summer of 1914 – and the price of that success, now that the war was over, was an endless succession of patients, and with them more and more doctors who wanted to be analysed and trained by Jung, so that they too could practise the new psychoanalysis.

So when Hermann Sigg suggested the trip to North Africa, Carl jumped at the chance. As he recounted to Emma in his letters, he and Sigg boarded the steamer at Marseilles, landing in Algiers and thence by train to Tunis, a journey of thirty hours followed by a further twenty-four to Sousse in central Tunisia. It was the first time Jung had come into contact with a civilisa-

tion utterly unlike his own. 'This Africa is incredible,' he wrote
to *Mein liebster Schatz*, 'my darling treasure', from the Grand
Hotel Sousse on 15 March 1920. 'Unfortunately I cannot write
coherently to you, for it is all too much.' Algiers was 'bright
houses and streets, dark green clumps of trees, tall palm crowns
rising above them. White burnooses, red fezzes, and among
these the yellow uniforms of the Tirailleurs d'Afrique, the red
of the Spahis, the Botanical gardens, and an enchanted tropical
forest . . .'. Carl the artist described 'A patch of deep blue sky, a
snow-white mosque dome; a shoemaker busily stitching away
at shoes in a small vaulted niche, with a hot, dazzling patch of
sunlight on the mat before him'. Back on the train he gazed out
of his carriage window hour after hour, mesmerised. 'Between
Algiers and Tunis lie 550 miles of African soil . . . whole Roman
cities, small flocks of black goats grazing around them, nearby
a Bedouin camp with black tents, camels, and donkeys.' At one
point the train ran down a camel, and always the deep blue sea,
glittering, and the olive groves and the palms: 'Then comes
Sousse, with white walls and towers, the harbour below; beyond
the harbour wall the deep blue sea, and in the port lies the sailing
ship with two lateen sails which I once painted!!!!' So there it
was, his dream from long, long ago, when he had painted his
longed-for vision of a town, with a tower by the sea and a
harbour with a sailing ship. And now he had the tower, and the
sailing ship, and everything.

Two days later he wrote again from the Grand Hotel Sousse,
having 'blown the desert sand off my table, which stands in the
columned courtyard of an Arab house'. He had been woken
early that morning by a great commotion. 'At the crack of dawn
the grunting and groaning of camels, many running footsteps,

sheep bleating, men shouting, then muffled drumbeats, the sky grows red in the East.' He got up and looked out of the window at the fabulous scene unfolding below:

> The square in front of the Hotel full of sitting camels, great numbers of black bearded faces in snow-white burnooses, shouts echoing back and forth . . . Three fellows with tremendous drums are drumming incessantly, along with them a sort of clarinet – flute-oboe, sounds like bagpipes – rapid rhythm, some men dance with arms outstretched – a caravan from the Sahara has arrived, about 150 men with many camels, to do one day of holy work for the Marabout, the holy man, who lives here – he feeds the poor from his garden – now the sun is rising majestically from the red and purple haze of the vast desert – three large flags of green silk with golden crescents unfurl . . .

He ended, breathless: 'I am having a shamelessly good time. I hope all is well with you, too. With loving greetings, your Carl' – adding by way of a postscript: 'I have been collecting some desert flora for Aggi.'

'At last I was where I longed to be: in a non-European country where no European language was spoken and no Christian conceptions prevailed,' he wrote in *Memories, Dreams, Reflections*. 'Where a different race lived and a different historical tradition and philosophy had set its stamp upon the face of the crowd.' He sat for hours in an Arab coffee house listening to the conversations, understanding not a word but studying the men's gestures and expressions of emotion. He reminded himself that this strip of land had already borne the brunt of three civilisations: Carthaginian, Roman and Christian. He could smell the blood. He wondered what effect the coming of Western technology would have on Islam.

After Sousse he and Sigg travelled further south to the oasis city of Tozeur. 'Towering date palms form a green, shady roof overhead, under which peach, apricot, and fig trees flourished, and beneath these alfalfa of an unbelievable green. Several king-fishers, shining like jewels, flitted through the foliage. In the comparative coolness of this green shade strolled figures clad in white, among them a great number of affectionate couples hold-ing one another in close embrace – obviously homosexual friendships.' The women were heavily veiled. If not, they were prostitutes, Jung's dragoman had explained to him. All this he relayed to Emma, back at home in Kusnacht.

Over the next few years there was a shift in Emma's life as her three eldest children grew up and left home. In September 1923 Agathe and Kurt were married at Seestrasse, without fuss – Agathe, beautiful in her wedding gown of cream silk and veil, Lil and Marianne, now nine and thirteen, their bridesmaids. But no fancy church wedding, just pastor Adolf Keller, Tina's husband, to officiate. Two years later Gret married Fritz Baumann, another ETH student – furious that her wedding day had to be postponed from an auspicious date in March, accord-ing to her astrological charts, to an average day in April, just because her father was away on his travels again.

Agathe had to leave almost immediately for Augsburg in Germany where Kurt had an engineering job, and by the time Gret got married Carl and Emma were already grandparents. In

all, Agathe and Kurt had three children, two girls and a boy.
Gret and Fritz had five, all boys. Franz was still at home but
having difficulties, unsure what to do with his life. His father
was no help, unwilling to offer any guidance, expecting him to
find his own way. Confused, he started to study medicine but it
did not suit him and he failed his exams. His impatient father
gave him some money and told him to go off travelling until he
found something he could do. It was Groma Rauschenbach
who, typically, came to the rescue and suggested architecture.
Franz ended up in Stuttgart with Ernst Fichter, the relation who
designed Seestrasse, and an architect he became. When Fowler
McCormick came back on a visit to Zürich after studying in
America he saw for himself how it was with Carl: 'I remember
him saying way back that really, if a family is healthy, the children grow up by themselves, you don't need to do much about
the children then. Of course I always thought it was fine to say
that, but Mrs Jung did a lot of valuable work with the children,
there is no doubt about that at all. They did not grow up just by
themselves. Most of the credit for the condition of the children
should go to Mrs Jung.'

With Carl frequently away, and just Lil and Marianne at
home, Emma had more time to pursue her own interests.
Throughout the 1920s she began to shape her lifelong research
into the Grail legend into a book. The twelfth-century romance
by Chrétien de Troyes about Perceval, a knight in the Arthurian
legends, which had fascinated her since her days in Paris, ran
like a thread through her own life: Perceval is born in a forest,
in twilight and only half conscious, protected from all dangers
by his mother. His father is absent, killed in battle. One day he
sees five knights in the forest, and he leaves his mother and

childhood behind to follow them. It is his odyssey, and each adventure brings him further into the world of consciousness. Finding the castle of the Grail he meets the Fisher King and embarks on his quest for the Holy Grail, described as a cup or vessel possessed of symbolic and magical powers – in later Christian texts transformed into the Holy Chalice which Joseph of Arimathea was said to have brought to Britain after the Last Supper, containing the blood of Jesus from the Crucifixion

The book, *Die Graalslegende in psychologischer Sicht* (*The Legend of the Grail from a Psychological Standpoint*), was a scholarly work with an extensive bibliography, ranging from classical derivations and Eastern parallels to etymology. 'The Grail Legend is an especially stimulating subject for psychological consideration because it contains so many features that are also found in myths and fairy-tales,' Emma writes in her introduction. The story was known to everyone, she explained: a mysterious vessel guarded by a king who is ailing in his castle – a place which is difficult to find. The king can only be restored to health if a knight 'of conspicuous excellence' finds the castle and then asks a certain question. This is the quest with its symbolic meaning: the quest of every individual for psychic health and wholeness achieved by asking the right questions and freeing themselves from the dark forces of the unconscious, symbolised by the forest where Perceval was born and spent his childhood. 'Moreover, it has lost far less fascination for contemporary men and women than have the latter [fairy-tales] which may indicate that it still embodies a living myth.'

Respecting Emma's prior interest in the Grail, Carl left the subject to her, though he himself had had a profound interest in it from an early age. It was one of the first things he and Emma shared when they were courting, talking like-minded as they walked high on the slopes above Ölberg to 'their' bench at the edge of the wood. And in fact Carl is never far away in Emma's text. His ideas on symbolism run right through it, above all Perceval's odyssey and his individuation from unconscious to consciousness. Every now and again Emma interprets a passage with specific, if unnamed, reference to Carl: 'a man naturally has the tendency to identify with his masculinity, and, as is well known, the acceptance of his feminine side is a severe problem for him. He is therefore inclined to act unjustly towards the feminine. Historically the woman was stigmatised as the devil,' and this followed shortly by: 'It often happens that people who behave in a markedly egocentric way are basically in constant danger of being absorbed by other people or situations, so that in such cases the egocentricity may be looked upon as a kind of bulwark against this tendency.' It is as though Emma, grappling with the labyrinth of her life with Carl, trying to understand him, uses the book as her own quest. She continued to work on the book for the rest of her life, leaving it unfinished when she died. It was Carl who made sure it was completed after her death, by one of their colleagues, Marie-Louise von Franz, and Carl who had it published.

These days Emma also had more time with her circle of female friends – Martha Sigg, Johanna Meier, Tina Keller, Susi Trüb. And Toni Wolff, always Toni. They spent evenings together, often discussing women's matters: their problems, their place in society, the social changes in Switzerland as a result of the war, their work, their hopes, their ambitions. And Emma was still the president of the Psychological Club, more estab-lished now after her diplomatic manoeuvres, though as troubled as ever with rivalries and jealousies. But now, finally feeling confident enough, and encouraged by Carl, she took the decisive step to train as an analyst herself, with their friend Hans Trüb. From Carl's point of view it was a perfect solution. Every time he went away he would be able to hand over his patients to Emma. Or to Toni Wolff, who was already training to be an analyst. A *ménage à trois* of a different sort.

The other *ménage à trois* carried on as before, the two women sharing Carl more or less equably these days, each with their own role. Emma at Seestrasse represented stable, everyday life. 'Her influence brought him every moment back to the normality of a normal Swiss. Whenever he was with her, he was just a normal Mr Jung living in Kusnacht and was in no danger of running away with his fantasies,' recalled Heinrich Fierz, the son of good friends of Emma and Carl. 'This is a very important function of a wife, and Toni Wolff had, of course, the function of the anima figure. It is always a fault [mistake] if the wife tries to play the role of the anima. This gives a terrific cocktail.' Carl Meier's wife Johanna remembered it in much the same way: 'It wasn't just a triangle situation like it is for most people.' At one point Emma and Toni shared an analysis with Meier, but nothing changed. Because there was no getting away from it: Carl needed both of them, for his sanity – Emma wife of Personality No. 1, Toni wife of Personality No. 2. 'Mrs Jung was more extroverted than Toni Wolff,' said Johanna. 'She could sit after lunch in the afternoon with her comic magazine . . . and just sit there and laugh and laugh. She had a great sense of humour. Both were strong women, but it was easier with Emma. 'She was his partner,' said Tina. 'They were equals. She could stand up to him.' And they could laugh together, as they always had.

By September 1920 Jung was in Cornwall for a series of talks organised by Constance Long. She had been introduced to Jung by her friends Dr David Eder and his wife Edith who had gone to Zurich before the war to be analysed by him. Long was one of a handful of impressive female doctors, English and American, including Beatrice Hinkle, who gathered round Jung. In Cornwall, about a dozen people attended the two-week session, everyone staying in the same boarding house: the Eders, Maurice Nicoll, Peter Baynes, an English doctor who had served with the Army Medical Service during the war and was now turning to Jung and psychoanalysis, and Esther Harding, who had graduated from the London School of Medicine for Women in 1914 and was a friend of Constance Long. This time Carl took both Emma and Toni with him. But at the end of the fortnight Toni went to London alone, and thence back to Zürich, whilst Carl and Emma travelled to Glastonbury, where Emma wanted to pursue her researches into the Holy Grail.

One day at Easter, 1922, a boy was standing on the shore of Lake Zürich, at the upper, isolated end, by a village called Bollingen. Hans Kuhn was twelve and lived in the village, no more than a hamlet, with his parents and siblings. As he stood there throwing sticks and skimming stones he saw a boat, a fine yawl with red sails containing a man and a woman and two children, making for the shore, and as the man steered it to the shore he called out to the boy to lend a hand. This was Herr Doktor Jung, it turned out, and he wanted to know if there was land to be bought along that shore. He had already tried to buy up one of the small islands on the upper lake where he often took his children camping, but the owner was not prepared to sell. Hans could have told the Herr Doktor that a purchase at Bollingen would not be popular either, it being the village bathing place, but he went to fetch his father as bidden. The Herr Doktor, a man who seemed used to getting his own way, came back at Whitsun and managed to make a deal with the owner and by the autumn he had engaged the mason from Schmerikon, who owned the quarry on the opposite shore of the lake, to ferry stones across and start building what would become the famous Tower at Bollingen.

Bollingen Tower – Carl's spiritual home.

The property consisted of a stretch of shore with a large meadow to one side. At first Carl wanted to build the tower himself, with help from the Niehus and Preiswerk boys and Franz, arriving by boat and camping in the tents he had brought back from Château d'Oex at the end of the war. But the stones were massive and the work too hard, even for him, and he was soon obliged to ask the mason from Schmerikon to engage some men to do the heavy work. Still Carl helped with the stones and oversaw every last detail, just as he had with the building of 228 Seestrasse.

The design was for a simple single-storeyed round tower in the medieval style, with a hearth at the centre and bunks for sleeping round the edge, and small windows in the thick walls piercing the darkness, creating a powerful, spiritual atmosphere, just like the towers in his dreams. Emma and the girls stayed away until the weather got warmer but then they came too, bringing provisions to cook on the open fire. There was no running water and no lighting other than oil lamps. Hans Kuhn brought milk from the farm every morning and water from the village well, and bread. It was a primitive life, exactly as Carl liked it, and an adventure. Soon he had dug a garden at the far side of the tower where he planted potatoes and maize, a favourite of his for breakfast. He did most of the cooking, because he liked it, and he liked it just so: if someone was helping him and they got it wrong he shouted at them, so mostly they just let him get on with it, dressed in his old gardener's apron and shorts and his battered straw hat.

After a big breakfast, it was bread and a piece of sausage or cheese in the middle of the day and a big meal in the evening, sitting round the hearth inside the tower if the weather was bad or outside round the campfire if it was good, Carl smoking his pipe and telling stories late into the night. Sometimes he would go into the village for a cheese and onion tart; once, when the tart wasn't to his liking, he swore at the baker and threw it in the air so it hit the ceiling. 'Odd, that Herr Doktor', is how the villagers described him. But once they got talking to him they usually changed their minds. He was so full of information and spellbinding stories, so interested in their lives, as happy to spend time with them as with the 'grand people' at Kusnacht. He was simple in his manner, often arriving at the village on his bicycle with a knapsack on his back, and content to spend long evenings playing games of *Zweierli*, two-handed cards, with Hans: at heart a peasant just like them. Later, when he was abroad, he wrote long letters to Hans, which the boy read out loud to his family, and thence to the whole village – marvellous letters full of unimaginable things.

As the tower grew larger and life there became more routine, Carl liked to sail to Schmerikon to visit the wine merchant and sit on his cellar steps tasting a good local Riesling or a fine Italian Chianti. Often he went to Bollingen on his own for days at a time, either sailing there or taking the local train and walking the rest, rucksack on back. Storms could blow up suddenly on the lake, and then the boats had to be battened down and the heavy doors of the tower bolted, and he would stay inside, reading and writing. But in summer Carl lived and worked mostly out of doors, sitting for hours at a time on the shore, digging his channels and dams, thinking things out, giving full rein to his unconscious. Often he was joined by one of his two 'wives'. When Emma came she set herself up at a small table a short distance from the tower to read or do her needlework. Toni Wolff didn't know how to cook and she didn't like camping or any kind of physical discomfort, so she didn't come to the tower till it had an upper floor with proper sleeping quarters. But then she too came, sometimes with Emma as a threesome, mostly alone with Carl.

In February 1923, shortly after the Jungs had bought the land at Bollingen, Carl's mother died unexpectedly after a short illness. It was from Emilie that Carl had inherited his sixth sense, his visions and voices and a respect for the occult which never left him. Her death brought up all sorts of spookery from his unconscious, though he recovered sooner than he would have done during his years of crisis. Ever since the end of the war Carl's inner life had steadied. Bollingen, the true home of Jung's Personality No. 2, felt like fate, a favourite word of Jung's. 'Gradually, through my scientific work, I was able to put my fantasies and the contents of the unconscious on a solid footing,' he wrote. 'Words and paper, however, did not seem real enough to me; something more was needed. I had to achieve a kind of representation in stone of my innermost thoughts and of the knowledge I had acquired. Or, to put it another way, I had to make a confession of faith in stone. That was the beginning of the "Tower", the house which I built for myself at Bollingen.' Stone had always held an almost magical power for Jung – the boy who sat on the stone in his garden wondering whether he was the stone or the stone was him. Bollingen was 'for myself' and no one else.

Carl's wanderings continued. It was Fowler McCormick who first suggested a trip to East Africa in the summer of 1925, an 'Expedition into Interior Africa', as the *Milwaukee Journal* would describe it, for research into 'the psychology of savage peoples'. Carl was only too happy to leave the house and patients at Seestrasse to Emma and Toni Wolff. The previous autumn he had gone to America at the urgent request of Fowler's parents, Harold and Edith McCormick. They had been divorced not long after their return to America after the war and Harold had remarried, but now they joined forces, begging Carl to come and give Fowler a good talking to. Fowler, their only son, was threatening to marry Anne Stillman, known as Fifi, the mother of his room mate at Princeton. Carl did his best, but to no avail. Instead he and Fowler took a long trip round America and New Mexico, including the Grand Canyon, Arizona, New Orleans, Washington, where Jung studied 'Negro patients' at St Elizabeth's Hospital, and Taos for two weeks, where Jung met Native Americans, in particular Ochwiay Biano, or Mountain Lake, an elder of the Hopi tribe, who talked to him about their rituals and the mythology of the sun. Christmas was spent in Chicago with George Porter, heir to a vast railway fortune and old friend of the McCormicks. Left to her own devices, Emma took her younger children off to Schaffhausen to spend Christmas with her mother in the comfort of Ölberg, as much a pleasure for her as for the children. Toni Wolff spent Christmas alone with her mother at Freiestrasse.

In the event Fowler did not accompany Jung to East Africa, but Jung's new friend, the Englishman Peter Baynes, did, plus George Beckwith, another wealthy American who was prepared to foot the bill. Unfortunately Baynes and Beckwith didn't get on, even though Baynes, who had moved to Zürich to become Jung's assistant, had been Beckwith's analyst off and on for some years. Matters were made infinitely worse by the fact that Baynes's wife, Hilda, had committed suicide in London on the eve of their departure. Baynes was guilt-ridden and morose throughout the journey. Beckwith turned out to be a champion sulker.

They sailed from Southampton for Mombasa in Kenya in October 1925. On board the ship was a group of young women travelling out to be married to officers in the Colonial Service. Among them was Miss Ruth Bailey, the sister of a bride-to-be. She was thirty-three years old and had been engaged herself, twice, but both men had been killed in the war. Miss Bailey hailed from Cheshire, one of those straightforward, adventurous, hearty and jovial Englishwomen from a good family who did not let life get her down. She did not meet Jung on board because he was not much in evidence, spending the six weeks avoiding his two travelling companions as much as possible, and learning Swahili from an old colonial officer. But they ended up at the same New Stanley Hotel in Nairobi, and during one of the fancy dress parties which the hotel laid on for their guests Ruth took herself off to a small drawing room to get away from all the noise and merriment and there she found Jung poring over a map of Kenya. He did not look up so she just sat there, passing the time. After about an hour he asked if she was interested in maps. She was. The following day he invited her to join them for breakfast, then a shopping trip to the bazaar. There was nothing sexual about it, she noted, he merely wanted a buffer between himself and his two awkward companions. Miss Bailey was underestimating herself. Jung immediately enjoyed her no-nonsense, direct and humorous ways and especially the fact that she had absolutely no interest in psychoanalysis. By the time the three men set off on their expedition, Ruth was on her way to Turbo in the Rift Valley Province with her sister and future brother-in-law, rather dreading the boring months of expatriate life which lay ahead. Jung had probably already made up his mind to ask her to join them.

Two months later a runner turned up at the bungalow in Turbo with an invitation for Ruth to come to Mount Elgon on the border region between Kenya and Uganda. The three men had embarked on what they called their 'Bugishu Psychological Expedition', with a large contingent of bearers and plenty of supplies, but they had had a tricky time getting there, as Francis Daniel Hislop, a junior colonial officer in a remote up-country station, discovered when he saw a large safari box-body car pulled up on the side of the track near his bungalow in the Nandi District. 'Good afternoon,' he said. 'Can I help you in any way?' To Hislop's surprise one of the men turned out to be Dr Jung of Zürich. Jung was delighted that young Hislop had heard of him. They wanted to get to Mount Elgon, he explained over tea in Hislop's bungalow. Hislop had to inform them they were on the wrong road and he wondered how they were going to communicate with the Karamojong and the Sabei, who did not in fact speak Swahili.

By the time Ruth Bailey arrived from Turbo, resisting all attempts to stop her embarking on this foolish and dangerous escapade, 250 miles of it, the three men had found Mount Elgon and Jung had found a man amongst the bearers who could communicate with the elders of the tribe, and spoke some pidgin English too. Fascinated by the three white men, the tribal people were happy to talk about their rituals and beliefs, and especially their dreams, to which they attached profound meanings, just as the white doctor did.

The postal service was hit and miss in those regions, but Carl managed to write to Emma his usual vivid accounts, and even sent a long letter to young Hans Kuhn. A snake had attacked Herr Beckwith, he wrote, and one evening he had been infected by sand-fly fever and dreamt that he was 'going black'. They would be home by the beginning of April. Franz recalled that this was the one and only time he ever witnessed a jealous scene between his mother and Toni Wolff: they fought over who should go and meet Carl at Marseilles. His mother settled it: neither of them. Franz, now eighteen, would go.

Perhaps Emma was anxious when she heard about the intrepid young Englishwoman who acted as a buffer between the three men and was able to make Carl laugh out loud. She need not have worried. Not long after Carl's return, Emma wrote to Miss Bailey inviting her to stay at Seestrasse. When Ruth arrived she found Emma sitting on the landing of the stairs having her portrait painted. Apparently George Beckwith, who adored Emma, had commissioned the painting as a gift for Carl, but the artist was having trouble making her smile. 'I remember this man saying to me "Come Miss Bailey, and amuse her. Tell her funny stories. She is much too serious. Much too serious." So Ruth quickly settled into her usual role and soon had Emma laughing and giggling like a schoolgirl at her stories of Kenya. Emma was beautiful, Ruth noted, but not happy. 'Her eyes were very sad,' she remembered, 'and I sat on the stairs below telling her all these daft things about Africa.'

Ruth hardly saw Jung except at mealtimes, spending all her time with Emma and the children. They went for walks, stayed with Emma's mother in Schaffhausen and her sister Marguerite living nearby, visited the galleries and theatres in Zürich, including a picture house where they watched an English comedy which made Emma laugh until she cried. Another time they went to a fancy dress party at the Psychological Club, to which Toni Wolff turned up as Queen Nefertiti, chain-smoking her cigarettes in her long cigarette holder. But though Ruth and Emma talked about many things, they never discussed Emma's marriage. After that first visit Ruth came to Kusnacht every summer, and the Jung girls went to stay with Ruth in Cheshire, as did Emma. After Emma's death Ruth was the one to come to look after Carl, staying with him till the end, as friend and carer, loved by the whole family.

Why was Emma so 'sad' in 1925? A clue lies in something Jung wrote about his obsessive travels at that time: to his own astonishment he realised it was partly to avoid 'personal problems', and the suspicion dawned on him that he had undertaken his African adventure 'with the secret purpose of escaping from Europe and its complex of problems'. He admitted to himself that this was not so much 'an objective scientific project as an intensely personal one, and that any attempt to go deeper into it touched every possible sore spot in my own psychology'. The atmosphere had become too 'highly charged' at home. In *Memories, Dreams, Reflections* he uses the word 'Europe' not 'home', which only appears in the 'Protocols'. But the sense of it is clear; juggling his life between two women was difficult. There were rumours of another 'infatuation' too. And his mother's death had brought up all sorts of buried memories. Split Carl concluded: 'My European personality must under all circumstances be preserved intact.' Split, or wanting everything both ways? Jung by his own admission was a trickster: 'I can let myself be deceived from here to Tipperary when I don't want to recognise something, and yet at bottom I know quite well how matters really stand.' As Fowler McCormick put it: 'For Mrs Jung it was a rough time, for she was a very lovely woman, a gentle woman, a good mother, and she loved Dr Jung.'

* * *

By now Emma had handed over the role of president of the Psychological Club to Hans Trüb, and it was as Alphonse Maeder predicted: Jung and Hans soon fell out because Hans began challenging Jung's theories. Jung resigned from the club, taking Toni and Emma with him, though it was the last thing Emma wanted to do, but she felt she had to stand by Carl. She tried again and again to reconcile the two men but the rift could not be healed. Perhaps she suspected that it was not only a professional disagreement but also a certain jealousy on Jung's part: Emma had become very close to Hans and Susi Trüb. When life got too difficult at Seestrasse she would sometimes decamp to their holiday home in Ticino in southern Switzerland. Both the Trübs were distressed by the rift, but Susi understood that Emma 'had to be on her husband's side. But I only saw that later; and then I never blamed her for it, because I too had to support my husband.' It was doubly upsetting because the two women were very close: 'With my sister [Toni] I only related as a sister. The personal relation was with Emma,' said Susi. 'She was truly and completely a woman.' It was Susi who knew Emma had considered divorcing Carl three times. There was something innately brutal about Carl, she felt, certainly in his dealings with Hans. 'People just came into conflict around him.' Soon enough it was Hans Trüb who had resigned from the club, and the holy trinity of Emma, Carl and Toni, as some members liked to jest, returned.

From March to July 1925, shortly before his extended trip to East Africa, Jung gave a series of seminars at the Psychological Club in Zürich charting the development of his ideas from the early days of his cousin Helly's seances, to his work at the Burghölzli, to Freud, and on to the sinologist Richard Wilhelm, whom Jung had met in 1923 and who had now a profound influence on his thinking. Jung spoke in English at these seminars, many of the twenty-seven people who attended being either English or American, including Esther Harding, who he knew from England, and two impressive American doctors now training to be Jungian analysts, Kristine Mann and Eleanor Bertine. Their friend Cary Fink, another trained doctor, took the notes. She had come to Zürich with her daughter, Ximena, in 1923 after getting divorced, and was soon part of the Jung circle of friends. After the seminars Jung liked to repair to a local café, everyone vying to walk with him, sit next to him, talk to him, grab a piece of his magnetic self. Emma preferred to go home alone.

Over the next few years Emma gave several talks to the Psychological Club, some on the Legend of the Holy Grail, others on the Jungian theme of animus/anima, which were later published in a slim volume. These turned out to be highly personal and revealing, rather as her talk on 'Guilt' had been. With her usual modesty Emma admitted she could not claim a complete understanding of the terms and would therefore limit herself to an examination of the animus and anima in relation to the individual and to consciousness. She began with a brief definition: animus was the masculine principle, anima the feminine, a man having some feminine characteristics, a woman some masculine ones. Normally both were present to a degree, but often hidden, conditioned by latent sexual characteristics and a person's experiences in life with the opposite sex. Being unconscious these could interfere in an individual's life, sometimes in a disturbing, even destructive way. So they had to be understood.

For a young woman, a man of physical power could easily become an animus figure, she suggested: heroes of legend and so forth. For an older and more exacting woman a man who had accomplished things, or offered spiritual guidance, would be a more likely focus. Each could be problematic because of the danger of projecting the image on the man. Some women, the active, energetic, brave and forceful ones, already had the masculine aspect harmoniously integrated, but if it was not properly integrated it could overrun the feminine aspect and result in brutal, aggressive behaviour, including in the sexual domain. Then the animus could be destructive to the woman and her relationships, leading to depression or general dissatisfaction with life. It had something to do with the modern times they lived in: the women's movement, advances in birth control, technological advances, psychology's discovery of the unconscious, and the law which still gave the man greater privileges. The result was a lack of self-confidence in women and a resistance to becoming conscious. In many ways it was like a household budget, declared Emma, the practical wife and mother: women had a finite amount of psychic energy, and it had to be well spent.

The lines between professional and personal were often blurred in those early days of psychoanalysis. Even so, Emma sailed close to the wind, barely bothering to disguise references to her own life and marriage. Women often fell back on a projection of their hopes and ambitions, she suggested, but that could rarely be successful for long, especially if the woman was in a close relationship with the man in question. Then the incongruity between the idealised image and the image-bearer often became all too obvious: 'We become aware, to our great confusion and disappointment, that the man who seemed to embody our image does not correspond to it in the least, but continually behaves quite differently from the way we think he should. At first, perhaps, we try to deceive ourselves about this and often succeed relatively easily, thanks to an aptitude for effacing differences, which we owe to blurred powers of discrimination. Often we try with real cunning to make the man be what we think he ought to represent.' That was not easy for the man either, who already had his own problems discovering his 'anima'. 'When a man discovers the anima figure, image or human, he finds it fascinatingly attractive and hence it appears valuable.' Everyone in the audience at the Psychological Club must have known whose animus and anima Emma was talking about.

Jung put in his pennyworth too, writing his paper titled 'Marriage as a Psychological Relationship'. Setting out boldly, he stated: 'There is no such thing as a psychological relationship between two people who are in a state of unconsciousness.' The bond to the parents unconsciously influenced the choice of husband or wife in the first place, either positively or negatively, compensating for everything which was left unfulfilled in the parents' lives. In a typical marriage, one partner is usually the 'contained', the other the 'container'. The 'contained' partner was the simpler case, 'grounded on a positive relationship to the parents', whilst the more complex partner was hindered by a deep-seated unconscious tie to parents and 'burdened with hereditary traits that are sometimes very difficult to reconcile'. These people, 'having a certain tendency to dissociation, gener-ally have the capacity to split off irreconcilable traits of charac-ter for considerable periods, thus passing themselves off as much simpler than they are. Their partners can easily lose themselves in such a labyrinthine nature, sometimes in a not very agreeable way, since their sole occupation then consists in tracking the other through all the twists and turns of his character.'

It was rare to have a smooth, crisisless marriage, he said. It was not possible to become conscious without pain, and usually that transformation began in the second half of life when the passion of the early years gave way to duty and the feeling of a burden, like a vampire, creating disunity with the self. 'This disunity with oneself begets discontent, and since one is not conscious of the real state of things, one generally projects the reason for it upon one's partner. A critical atmosphere thus develops, the necessary prelude to conscious realisation.' It was at that very moment of crisis that everything could begin. This was one of Jung's most important and practical insights: crisis may feel intolerable at the time, but it was only through crisis that development of the personality occurred. As for the partner: she had to give up that *he* belonged to *her*, and stop looking for her security in him but find it in herself, at which point she became valuable again.

Did Emma manage it? It would seem so. In 1927 Jung was asked to lecture at the Kulturbund in Vienna. He took Emma with him, and they were met at the station by the vice-president of the association, Jolande Jacobi. 'I went to the train to fetch him and his wife and I met two wonderful people, looking marvellous, very alert, very jolly, and not at all [as expected].' Jacobi had evidently been forewarned. 'I was absolutely impressed when a beautiful tall, handsome man with beautiful white teeth, laughing, arrived with a charming woman with dark hair and light blue eyes. Like a saint, she looked, Mrs Jung.'

Jacobi was a thirty-eight-year-old good-looking, extroverted, charming Jewish Hungarian. She lived in an elegant apartment with her two young sons, her husband having returned to Budapest for unspecified reasons, possibly because his wife had a longstanding lover. Perhaps Jung hesitated before accepting the invitation to Vienna because of Freud; there had been no rapprochement between them since the final breakdown in their relationship in 1914. By all accounts Freud had not found a crown prince to replace him and he still maintained that Jung had broken away from him and betrayed him, rather than admitting that Jung might have been on a different tack from the start. Freud told Ernest Jones he would not be attending the lecture, pleased to miss 'an excellent opportunity of hearing about the Structure of the Soul from a first-rate source'. However, over a thousand people did turn up to the lecture, which was unheard of at the association, but it turned out to be a disappointing event, the shadow of Freud probably damping Jung's style.

Yet here they were in Vienna again, Emma and Carl, after all this time. They spent the next five days with Jolande Jacobi, who was as enchanted with Emma as she was with Carl. One evening they went to the theatre to see the musical comedy *No, No Nanette!*, during which Carl laughed 'so terribly loud and yelled so that the performance was stopped and the prima donnas greeted him with their hands, and he threw kisses, and they threw kisses, and there was quite an upheaval in the theatre'. Jacobi had never seen anything like it, but Emma was used to it. 'She was the most generous wife you can imagine,' recalled Jacobi. 'Mrs Jung was always tactful and wanted to leave her husband alone with the person he was interested in. She said she would like to see the town and left him alone with me.' The three became firm friends. There was no need for Emma to worry: sophisticated Jolande Jacobi was never going to fall in love with Jung. For one thing he had 'dirty fingers' and he was *schlampig*, a bit messy, whereas Jacobi had fine manners. 'He was somewhat heavy,' she concluded, 'a peasant-like Swiss who did not know how to deal with ladies of high society who flirted with him.' By the end of February they were home again.

In the winter of 1928 the Lake of Zürich froze over. People walked and sledged across it to visit friends, wrapped in furs and hats and muffs, quite dazzled by the whiteness and the magic. To her children's surprise, Emma strapped her skates onto her boots, a thing they had never seen, and took off into the ice-blue.

15

Coming Through

We were *jung* [young]
We were called Jung
To eternal youth
We belong

Lines carved in the terrace wall at Seestrasse

In 1929 Emma was forty-eight. Two daughters, Agathe and Gret, were married, and she had become a grandmother. Franz, twenty-one, had left home for Stuttgart to study architecture. Her two youngest daughters, Marianne and Lil, were nineteen and fifteen respectively. She felt more self-assured these days, though plenty of difficulties still lay ahead. But her central dilemma – how to manage life with her husband Carl and how to find her way through it all, the dilemma she had been grappling with ever since her marriage in 1903 – that was behind her. Or if not exactly behind her, then beside her, flowing along more like a stream than a torrent, winding its way towards the open sea. Perhaps most important of all: she was past the age of menopause, past the danger of having too many 'little blessings'.

She had taken her decisive step: to work as an analyst in her own right. In a sense she had been training for it ever since the early years at the Burghölzli, unwittingly at first. But as the years passed and she became increasingly involved in Carl's work, she found she knew more about it than she realised, not least because of living with split Carl. By the time of the Weimar Congress in 1911 she was acquainted with many of the leading lights of psychoanalysis, including Professor Freud of Vienna, and she had met some forward-thinking women too – Hedwig Bleuler and Beatrice Hinkle among them – who had introduced her to new and challenging ideas about women in society. Encouraged by Carl, Emma had become the first president of the Psychological Club of Zürich, where she discovered she could hold her own and rise above the jealousies and rivalries which somehow always surrounded her husband. Her own jealousies were harder to resolve, but now, helped by embarking on her own working life, she found she could contain them if not actually banish them.

As though to celebrate, Emma learnt to drive a motor car in 1929, taught by faithful Müller, who these days functioned as chauffeur and general handyman as well as gardener, and bought herself a Dodge Sedan. Carl learnt to drive at the same time, buying himself a red Chrysler Convertible, which he called the 'Red Darling' and which he parked outside Toni Wolff's house on Freiestrasse every Wednesday for all the world to see, like a red rag to a bull.

Emma's first analysand was the young Englishwoman Barbara Hannah, who later wrote the first biography of Jung. Hannah's father was Dean of Chichester Cathedral, and she had so far led the kind of life typical of people from her background:

Church of England, conventional middle class. But one way or another she had reached a point of crisis in her life. She came to Zürich hoping to be analysed by Herr Doktor Jung, but just then he was in a phase of trying to extricate himself from the demands of so many patients, all craving his attention. 'Oh my God, you bore me!' he announced, and promptly passed her over to Emma. Later she was passed on to Toni Wolff. That was how the threesome worked those days, passing patients on from one to another. When she came back to Jung he had changed his tune: 'Ah, you've ditched both of them, and now, you hellcat, you want to get me in the ditch,' he roared at her. Jung loved American slang. Hellcat was good. 'I knew you were a hellcat from the start. That's why I passed you on. But I tell you what I will do. I will take you on again on the understanding that, if you get me in the ditch too, it's up to you to get me out again.'

But Hannah started with Emma. 'She was just beginning. It was all very much like going to tea with a lovely and gracious lady, and only afterwards did one realise that she was an excellent analyst,' she recalled. 'She used to give you two hours instead of an hour. She did a tremendous amount for me.' Another analysand put Emma's method this way: 'She approached the problem you brought to her quietly, even tentatively, but there was no fumbling. To be "right" did not appear to interest her. She met you where you were at.'

At the Psychological Club, Emma was known as 'the unofficial authority of the Club', not because she was the wife of Jung but because of her 'strength and serenity'. When Carl offended people with his sudden rages or rudeness or sarcasm, Emma appeased them. When he became too boisterous playing games such as 'Alleluia' – where people threw a ball at one another shouting *Alleluia!* as a means of accessing their unconscious, and which could quickly descend into chaos – she kept him in check. She never joined the Wednesday lunches where he was surrounded by 'those fawning women, hanging on every word and fighting among themselves to get into his graces', as the forthright American Jane Wheelwright described, preferring to go home alone.

One afternoon in March 1931, Peter Baynes arrived at Seestrasse unannounced and in a state of high tension. Carl was away lecturing. Emma invited him in and they settled down to tea and talk. 'Emma was just lovely,' Baynes wrote to his future wife, Anne Leay, back in London. 'We just seemed to get to a calm leisurely mood like a broad river where we found the most human understanding. She told me lots of things about herself and her deep feeling conclusions seem to me ever so true. So you see I was able to tell her about you and me.' Baynes was in a difficult situation: since his return from the Africa trip, and in the wake of his wife's suicide, he had married – against Jung's advice – Cary Fink, newly divorced and living permanently in Zürich with her daughter. But the marriage was not working out, and now he had met and fallen in love with Anne in London and was stuck between the two women, unable to move forwards or backwards.

Emma calmed him. 'She got your spirit, as a woman I felt, and she said: "You see Peter, I would never say that the way things are in our lives [meaning Toni and C.G.] is in any way a solution . . ."' Emma was warning against a love triangle. 'She let me know how she had suffered and how she still suffers.' Her advice must have helped: Baynes married Anne that same year, and it turned out a happy and successful marriage which lasted for the rest of their lives.

In March 1932 Emma's mother Bertha Rauschenbach died. Emma was profoundly affected. She adored her mother and loved spending time in her company. Whenever she was getting lost in the labyrinthine problems of her marriage, it was in her mother she confided. After all, it was Bertha who had brought Carl and Emma together again after Emma refused Carl's proposal of marriage. Summer holidays, Christmases, weekends, so much of Jung family life had taken place at Ölberg. Now Marguerite and Ernst moved in to take over the house, but it wasn't the same.

Within months came word of another death. 'Dear Mr McCormick,' Emma wrote to Harold McCormick on 1 September 1932, having just heard of Edith's death. 'The news of Mrs McCormick's death was a great shock to me, and thinking of her brought up so many reminiscences that I must send you a word of sympathy.'

After the McCormicks' divorce Edith had retreated to her vast mansion on Lake Michigan, no longer taking part in the high life, in fact not taking part in anything very much at all. 'A tragic fate', as Emma put it. Harold had been a frequent visitor when Edith was ill, which Emma suggested must have been a great relief to Edith but perhaps also to Harold: 'In my thoughts she will always, as you too, be linked with a very intense and important phase in my life, and I shall always be grateful to have had the privilege of coming to know her and you, and of all the stimulating and precious contributions to my own development

I got through this relation. And of course the Club will always be a living and lasting expression of her great and generous personality, to which we owe it.'

Life at Seestrasse went on as usual, wrote Emma. Jung was at Bollingen, where he spent most of the summer holidays. 'I am also doing some analytical work and I like it very much indeed. I find it most satisfactory, and am sure it does me at least as good as the analysands.' She ended with the hope that everything was well with Harold and sending her warmest regards. 'Yours very sincerely, Emma Jung.'

Edith's death had taken Emma right back to the time, twenty years earlier, when her misery and crisis about her husband's 'infatuations' had forced her to confront her own demons. Few close to Emma and Carl in 1912 thought she would ever manage to overcome the difficulties, but gradually they had changed their minds. 'I think she underwent the most spectacular trans-formation during their married life. More than any of the women I saw. She was quite an exceptional person,' observed Carl Meier. His wife, Johanna, who was able to observe Carl and Emma at close quarters over the years, saw that not only did Emma find a way to manage her own life, but increasingly she underpinned Carl's too. 'I think this relationship, the real marriage relationship, was carrying his life work actually, certainly you could see this when you watched them,' she said. 'When you saw this couple it was a very harmonious couple. They belonged to each other. Certainly they did. I always had the feeling that his strength, both physically and spiritually, came through this marriage. He was in the centre and carried it all. But I'm convinced that this marriage carried the load, actu-ally, of the whole Jungian development. If this marriage hadn't been right and true, it would not have gone through the mill. I'm sure only Mrs Jung could do it.' They could see now that it was a 'joint career' in every sense, to the extent that the world would not have had the Carl Jung it knew without Emma Jung, steady in the background.

Johanna had once gone to talk to Emma about some difficulties in her own marriage. 'Well, do you want to carry on or are you going to quit it all?' Emma challenged her. Emma was no quitter evidently. 'I'm sure that each of them knew that it was worthwhile to carry through,' said Johanna. 'They belonged together.'

In September 1932 Emma was invited as guest of honour at the Analytical Psychology Club in London for a discussion of her paper on 'Animus'. She stayed with Peter and Anne Baynes, taking Franz and Lil with her. During her stay Emma and Anne became friends, later cemented by Anne travelling to Zürich to spend a month at Seestrasse being analysed by Emma. The analysis made everything extraordinarily clear, Anne said, noting a wisdom and warmth in Emma which she had not had from her own mother. 'I do love Emma. She does symbolise for me the kind of woman I want to be.' This was Emma's strength: to be simple and honest in her approach, empathetic but not directive, helping people to find their own way, as she had hers. Or, to use the Jungian term: to individuate.

And yet Toni Wolff was still there, still part of the *ménage à trois*, still claiming her share of Carl with her cleverness and spirituality, arriving at Seestrasse in her elegant outfits and big hats, puffing away at her cigarettes, wandering around the garden in deep conversation with Carl, irritating the children, causing Emma distress. 'Toni was a manipulator in a very lady-like way. She'd sure get rid of her enemies, but you never saw it happen. Emma was more forthright, less of a manipulator,' is how Jane Wheelwright remembered it. In January 1930 Carl went to Berlin to give a talk. He took Toni with him. As usual he left all the arrangements to his travelling companion. In a letter to a colleague Toni wrote asking for two single rooms in the hotel, but interconnecting, with a joint bathroom. In October 1935 he took Toni, not Emma, to London for a series of lectures he had been invited to give by the Institute of Medical Psychology at the Tavistock Clinic, introducing her to everyone as his 'colleague', though there can't have been anyone there who didn't know her true status – so unconventional, so un-Swiss, and how impressive was Miss Wolff's collection of stylish hats.

The Englishman Joseph Henderson, who first came to Zürich to be analysed by Jung in 1929 and later became a leading analyst, saw it this way: 'Jung told me personally once that the most valuable thing in his life was his marriage. He would give up anything before he would give up his marriage.' Toni Wolff knew it too. 'The thing went on for a lifetime because Toni was not married, and she was so alone and she had given her whole life to Jung,' thought Jane Wheelwright. 'She lived for him. And he could not dump her. So it was tough.' Barbara Hannah thought Toni Wolff learnt to accept it over the years: 'Toni overcame the besetting sin of so many single women, the desire to somehow destroy the marriage and marry the man herself.' In truth, Toni had little choice. 'The law Jung was faithful to was the creative instinct in him, he would abandon almost anything for that,' said Carl Meier. And Toni was still crucial to that. 'In their discussions she played an enormous role. She sometimes became quite dramatic too. Sometimes it became rough. Real rough.' Emma was no pushover either. 'She was a tremendously forceful person. She could put her foot down and attack you face to face and ask you to be really honest. She really was a forceful person, not someone who is easily pushed around.'

Toni's advice to a patient whose husband was having an affair was: ask the other woman to lunch.

> You would then get to know her a bit, you might even like her ... Sometimes if a man's wife is big enough to leap over the hurdle of self-pity, she may find that her supposed rival has even helped her marriage! This 'other woman' can sometimes help a man live out certain aspects of himself that his wife either can not fulfil or else she doesn't especially want to. As a result, some of the wife's energies are now freed for her own creative interests and development, often with the result that the marriage not only survives but emerges even stronger than before!

She made it sound easy, but it was not easy for her, still living with her mother and seeing little of Carl. But she had her work, and her analysands, who appreciated her cleverness. 'She did not give herself easily to a human relation, but once the ice was broken she was loyal and steadfast', said one analysand, Gerhard Adler. Another felt the much same: 'My original image of her was of a rather tall, very serious, erect, straight-backed woman – the way she always sat in her chair. Later I saw her as a warm, comforting, gentle, relating person.' Apart from her private practice, Toni Wolff headed the lecture committee at the Psychological Club, and for over forty years she assisted Jung with his 'group psychology' experiment, another of his forward-thinking ideas, designed to overcome the transference difficulties of one-to-one analysis. Work was Toni's salvation, and her constant link to Carl.

Each year, from 1933 onwards, Carl and Emma attended the annual week-long Eranos conference at Ascona on Lake Maggiore, organised by a wealthy Anglo-Dutch woman called Olga Froebe-Kapteyn, who aimed to bring together thinkers from many disciplines, including all forms of spiritual research such as the occult and theosophy. As usual people flocked around Jung, but these days Emma held her own. A photograph taken of the Jungs at one of the conferences shows a couple married for over thirty years, relaxed and at ease with one another: Emma in a white summer dress and hat, laughing, happy, looking directly at camera, Carl beside her, leaning towards her, pipe in hand, smiling, satisfied. When Jung was invited to America in 1936 for Harvard's tercentenary, to receive an honorary doctorate and to lecture on 'Factors Determining Human Behaviour', Emma went too. She attended all of Carl's lectures and seminars, and whilst he was seeing private patients, who came from as far away as California, she went sightseeing, probably with Fowler McCormick, who seems to have accompanied them for most of the trip.

After Boston they travelled to Bailey Island in Maine at the invitation of their new friends Eleanor Bertine, Esther Harding and Kristine Mann, where Jung gave another series of lectures and Emma was able to spend time with her old friend from the Weimar congress, Beatrice Hinkle. They stayed in Kristine

Mann's large old family house overlooking the sea and whenever there was some free time, Carl went sailing. Bailey Island
was connected to the mainland by a series of wonderfully engin-
eered bridges, but the coastline was untouched by the present
– most appealing to the Jungs. They left for home in October,
stopping off in London to catch up with more old friends and
for Jung to give a series of lectures at St Bartholomew's Hospital.
Barbara Hannah met them at Waterloo station and drove them
to a small hotel off Regent's Park where Jung again saw private
patients. By the time they got back to Kusnacht they were both
exhausted. Jung cancelled his seminars at the Psychological
Club and the lectures at the ETH to concentrate on the study of
his new enthusiasm: alchemy. Emma took up her usual pursuits:
the home and family, her research, her book, her analytical work
– and Carl.

By the spring of 1937 Carl and Emma were travelling again,
first to the Eranos conference, then back to America, this time
to Yale for Carl to give the Terry lectures on 'Psychology and
Religion'. The packed audience included Mary and Paul Mellon,
who would soon step into the Jungs' lives in much the same way
that Edith and Harold McCormick had done twenty years
earlier.

Like the McCormicks, the Mellons were fabulously rich, from
the Mellon banking family, and in time they became the founders
and funders of the Bollingen Press, publishing Jung's works in
English. Mary Mellon was lively and fun and a classy dresser, but
she suffered all her life from acute asthma attacks, and, like Edith
McCormick, she was seeking a cure. 'I was sitting directly
beneath the platform under a large black hat,' she recalled, 'and
I understood not one word, but I thought, though I don't know

what he means, this has something very much to do with me.'
When Carl and Emma went to New York the Mellons followed
to attend Jung's seminars, hoping to have a private session of
analysis with him. But Carl was too busy with the lectures and
seminars and another hectic round of social engagements, with
Esther Harding on hand to drive them about. Anyway, he and
Emma wanted some free time to visit the museums. Carl's lecture
at the Plaza Hotel was followed by a banquet in his honour. In
his impromptu speech, apparently still thinking of religion, he
said all we could do was follow Christ's example and live our
lives as fully as possible, even if it was based on a mistake. We
should go and make our mistakes, because there was no full life
without error. This was Jung at his practical best again, allowing
for mistakes, his own included. At another supper party the
guests, including the Mellons, were somewhat taken aback when
Jung took out his Swiss Army penknife to open a bottle of
Chianti, then retrieved a stick from his dinner jacket pocket and
started to do a bit of whittling, Swiss-peasant style.

Mary and Paul Mellon at Eranos, 1938, with Jung in the background.

Not everyone warmed to Jung's provocative manner, suspecting he put it on deliberately as a rejection of the decorum Americans expected from their honoured guests. He was too loud and brash, they said – he dominated dinner conversations, ignored people he did not want to talk to, and flirted with their women. Worse: there were rumours that Jung was pro-German if not actively pro-Nazi, perhaps anti-Semitic, based largely on those of his writings which addressed differences in racial and national psychologies, including 'Negroes' and American Indians, as well as Christians and Jews, and on the fact that he remained affiliated to the International Society for Analysts in Germany, accepting the editorship of their publication the *Zentralblatt* after their president, Ernst Kretschmer, a Jew, was forced to resign when the Nazis came to power. After the war these accusations were taken up again, mainly in America, quoting the article he had written for the *Zentralblatt* about racial differences, but instead of Jung's word 'differences', the word 'higher' was used, suggesting racial superiority. The rumours were also fanned by some of Sigmund Freud's more ardent followers, Ernest Jones amongst them, making it hard to distinguish fact from fiction. Freud, who had suffered true anti-Semitism all his life in Vienna and was soon to flee the city and the Nazis for London, surely knew the difference. Rumours flew about, 'cock-and-bull stories' as Jung would describe them: 'Whatever I touch and wherever I go I meet with this prejudice that I am a Nazi.' They were not persecution delusions, he said. They presented a very real difficulty.

Even with the benefit of hindsight, Jung does not come well out of this. But racial theories of the sort were common then, and no one yet knew what the Nazis were capable of. Emma's role during these visits to America was the usual one: as smoother and pacifier, keeping Carl steady. As to the Mellons, they travelled to Zürich and the Eranos conference at Ascona in order to have private sessions with Jung, which they did in 1938 and 1939, by which time another European war was once again threatening.

The Jungs arrived back in Seestrasse in late October 1938. By December Carl was already off again, this time to India with Fowler McCormick, stubbornly refusing to get inoculated before he left, trusting to fate. Emma and Toni stepped in to take over his analysands as usual. Toni went on organising the lecture committee at the Psychological Club, Emma went on giving her lectures on animus, and the Legend of the Holy Grail, and continued writing her book, her life's work, as close as she could ever get to Carl's complex inner world. When she was not working she liked to stay with Agathe and her growing family in Baden, or with Franz in Stuttgart, or Hans and Susi Trüb in Ticino.

Carl came back from India delighted to have found that in Eastern thought the concept of evil went hand in hand with the concept of good, something he had discovered for himself as a schoolboy in his vision of the turd falling on Basel cathedral. He also came back with amoebic dysentery. But it did not stop him travelling to England with Emma the following Easter, 1939. They stayed with the Bayneses again, the main purpose of the trip being for Emma to visit, or revisit, the sites connected with the Holy Grail: Glastonbury, Stonehenge, Avebury, Cerne Abbas. 'It was an extremely happy time for all of us, I think, for Emma was doing what she most wanted to do. Everybody was extremely carefree, somehow,' recalled Anne Baynes. Jung was in one of his high-jinks moods, making everyone laugh, bemusing guests in their Glastonbury hotel. 'Other people in the hotel just could not make out how four people could rock with laughter every morning at breakfast, because it isn't an Englishman's best time for laughter, as a rule. They were all having their quiet breakfasts' – shattered, no doubt, by Jung's bellows. As they drove into Salisbury they saw newspaper billboards announcing that Germany had invaded Czechoslovakia.

By the late 1930s Carl and Emma's 'joint career' was well established. He discussed all his ideas with her. She read everything he wrote. She looked after his patients when he was away, and she picked up the pieces he left in his wake. She kept him steady during his times of doubt or incipient insanity. Finally, Carl told Emma about his early traumas and his phallus dream – the great secret.

'My entire youth can be understood in terms of this secret,' he wrote. 'It induced in me an almost unendurable loneliness. My one great achievement during those years was that I resisted talking about it with anyone. Thus the pattern of my relationship to the world was already prefigured.' When he heard his pastor uncles discussing theology he thought: 'Yes, yes, that is all very well. But what about the secret?' His only consolation was the stone – the leitmotif to his secret life. 'Somehow it would free me of all my doubts. Whenever I thought that I was the stone, the conflict ceased.'

Perhaps the stone was bound to lead Jung to alchemy sooner or later. The sinologist Richard Wilhelm had sent him a Taoist alchemical text, *The Secret of the Golden Flower*, which Jung described as 'the first event which broke through my isolation. I became aware of an affinity; I could establish ties with something and someone.' Soon he was deep into the symbols and secrets of alchemy – another route to deciphering the unconscious.

But alchemy was a step too far for Toni Wolff. From the mid-1930s their relationship was on the wane, until finally it almost ceased. Most observers thought the break came from Carl. They saw how hard it was for Toni. She had met Carl when she was twenty. As her sister Susi said, she had no other partner but Jung all her life. 'She was afraid. She was a very lonely person, and in Jung she surely found the man or partner, someone who could correspond to her intellectual [spiritual] nature. And so that was the way for her, I think. This was everything for her. She had some connexions to other people; she was a very good analyst. But somehow she was never totally in life.' She still came to Seestrasse on Sundays, but it wasn't the same. Ruth Bailey, visiting, noticed how Carl would take out a book and start to read when they were all sitting on the terrace having tea. 'That was a terrible situation and I used to think: "Why do you ever come? I wouldn't come if I were you."' Reflecting on it years later Franz Jung said: 'When I was a ripe man, I felt sorry for Toni. She never knew what it was to be a wife and mother.' In contrast to his own mother who, he said, 'was sure of herself and she was sure of her femininity, so she could deal with Toni.'

Fowler McCormick, commuting between America and Zürich, noticed that Emma and Carl's relationship was getting stronger whilst the relationship with Toni became more distant, until finally in later years Carl could be openly quite cutting and hurtful to her. Often now he left the two women, Emma and

Toni, together, and took himself off to his *Cabinet* until it was time for Toni to leave. Even Emma, after all the years of tolerance, was losing her patience. Once when Jolande Jacobi was visiting from Vienna, Toni Wolff arrived and Jacobi made to leave. 'No,' protested Emma. 'This woman tries all the time to mix herself in my house. Every Sunday she must have my husband. No, you are our guest and you stay.' It was quite unlike Emma, certainly unlike the young Emma. But these days she was less accommodating. 'Well, she was a marvellous woman,' concluded Jacobi about Emma. 'She carried on during this friendship of her husband with serenity and with enormous generosity, until she won.'

When the Second World War broke out Franz and all Emma's sons-in-law – three of them as Marianne had meanwhile married – were called up for border control duties. Jung was too old, but he was drafted as a doctor and psychiatrist for the district on the west bank of Lake Zürich. The garden was once again dug up by Müller to plant potatoes and vegetables. Mary Mellon and other American friends wrote to ask how they could help, but there was nothing to be done except send food parcels. Mary sent coffee, sugar, even butter, via Macy's and the Red Cross. 'I think the night has descended upon Europe,' wrote Jung. Gret arrived from Paris, where her husband worked for a Swiss firm, bringing her five sons to live with her on the top floor of 228 Seestrasse. After the fall of France, Jung moved the entire family up to the Bernese Oberland for safety, including Franz and his new wife, Lilly, expecting their first child. Emma had wanted to leave Kusnacht for some time, but Carl had been reluctant to abandon his medical post. Then came a telephone call from an official in Bern warning Jung that he was on a Nazi 'blacklist', no doubt for his 'degenerate' writings, and would be arrested if Germany invaded Switzerland. Lilly gave birth while they were up in the Alps. But once it became clear the Germans would not invade, the family returned to Seestrasse.

Emma noticed that Carl looked exhausted, quite unlike his usual ebullient self. In fact he had not been fully well since his return from India in 1939. His work as a doctor also tired him. He looked thinner and older. In February 1944 he had a heart attack. He was already in hospital when it happened, in a private clinic near the Burghölzli, where he was being treated for a broken fibula from a fall while out walking in the snow. The leg was set and he was ordered to stay in bed, which caused embolisms in his lungs and heart, in turn causing a cardiac arrest. He remained in hospital until the end of June, and in all that time Emma never left his side. She took a room in the same corridor and rarely left the clinic, getting her daughters and daughter-in-law to bring in whatever was needed. Visitors queued up to see the patient but they were often refused, even Toni Wolff. Usually it was at Carl's bidding, but it was Emma who enforced it. Susi Trüb remembered how shocked Emma was that people who were refused entry could be so rude. They seemed to think Carl belonged to them. Susi also noticed how close Emma and Carl were, or perhaps how clear it now became that they had been all along.

By the time Carl came back home he was so weak he could not climb the stairs and had to have a daybed in the *Stube*. Gradually Jung started to walk round the room, then the garden. By the end of the summer he was almost well again, but Emma could not let her new role go, overseeing him, constantly checking for any dangerous signs. It did not always go down well.

Ruth Bailey saw how things were. 'I think one time she would almost have had him in a wheelchair as an invalid, so as to care for him, to make quite sure. She was not possessive, as such. She wanted to keep him going, keep him well, you see, and of course, he was older than she was.'

Once Jung was better, the grandchildren came and went again. The Baumanns had moved to a house in Kusnacht, but the five boys were always biking or running up Seestrasse, down the yew-lined drive and into the house. It was open house at Seestrasse, their grandmother always welcoming, ready with her bowl of Sprüngli chocolates, and often they bumped into other cousins already there. They swam, they boated, they chased around the garden, or drew, or played cards at the table in the *Stube* if it was raining. In fact, they could do pretty much as they wanted, just as the Jung children had. Their grandfather was not often around. He was either in his *Cabinet* working, or at Bollingen, or off giving a lecture somewhere. But when he was around it was wonderful. He told them stories, he took them sailing, he lost his temper, he swore like a Swiss peasant: *Hoore Verdammt! Du Loeli!* You bloody idiot! He laughed his loud laugh. And he cheated at games. But then it was time for him to go back to his work. It was their grandmother who was always there and made time for them, as much as they wanted. She never forgot a birthday, and she had a talent – they all agree – for finding just the right present. Christmas followed much the same pattern as it had when their parents were young: crackers and games and entertainments round the piano, and the tree in the corner lit up with candles and Christmas hearts and stars. When Adrian Baumann, Gret's son, was ten he asked to see his grandmother's jewels. She took him up to her room and showed them to him, one by one, each with its own story. He never forgot it. 'She was the heart of the family,' he said. 'We adored her.'

Emma and Carl had nineteen grandchildren, and as the children grew older they began to understand how it was with their grandparents. They saw how strong their grandmother was underneath her quiet exterior: never frightened, as others were, of criticising their grandfather if she thought he had said or done something wrong. 'She was strong,' says Adrian, 'but very feminine.' She was always modest, never showy, recalls his brother Dieter, the eldest grandchild, but she had great influence on their grandfather, moderating him: 'It's remarkable how she was able to steady him.' Emma was clever, thought granddaughter Brigitte, Agathe's daughter. And 'always the lady'. It was deep in Emma, this modest grandeur. 'Oh she was la Grande Dame,' one of the younger members of the Psychological Club recalled. 'She was very simple, absolutely Swiss, she was the great Bourgeois Swiss, you know. But she had something more than this. She had something absolutely superior, *une grandeuse mere*. She was the only person there [at the club] who had really individuated.' Dieter agreed: 'There was something aristocratic about her, but in a positive way.'

It was with her children and grandchildren that Emma could best be herself. 'She treated us all the same, as she did her own children. My mother really adored her,' says Adrian. She was unusually open-minded. You could talk to her about anything. But unlike their grandfather, she could not let out her feelings. That was her Achilles heel. She had deep feelings, but she could not show them. Their grandfather, on the other hand, showed everything, good and bad. In the Switzerland of those days, grandparents greeted their families with a handshake. Lil's son, Hans, remembered how warm his grandfather's handshake was. And 'the totality of his laugh'. Brigitte remembers his Basel sarcasm and being a bit frightened of him. 'Sometimes he was very loud and sarcastic, while she was very quiet,' said Jost Hoerni, another of Lil's sons. 'We loved her, really, because she was very nice with us and she spoiled us a little bit, as grandmothers usually do. She was a very different temperament than grandfather.' Andreas Jung, Franz's son who took over the Jung house with his wife Vreni when their children were small, remembers his grandfather's large personality. 'He was so powerful,' agrees Dieter.

On Sunday afternoon walks it was clear there was no tension between their grandparents. 'Children can feel such things,' says Adrian. 'Even in Bollingen, I never felt any tension. Of course he had a temper, but it never lasted.' When he was a child he did not understand about his grandfather's flirtations, but he is sure it made no difference, at least to Carl. 'It didn't mean they didn't have a good marriage. I'm absolutely convinced that he really loved her. Even with those escapades. But it must have been difficult for her. Toni Wolff. I don't know any woman who could bear it. But I had the feeling she always loved him. I'm sure she did.' His brother Klaus agrees. 'He depended on her more than she on him.' He remembers his grandfather as a bit of a joker and a cheat. 'There was something in his trickiness which attracted her to him.' They all remember how much they laughed together. 'Oh yes! Of course! He had a great sense of humour. And so did she.' Dieter, now in his late eighties and an analyst himself, says his grandmother was aware of how much she owed their grandfather, who always encouraged her to do her own work, find her own way, individuate. 'Both learned from each other all their lives,' he says. 'They are a great example of this.'

Toni still visited on Sundays. As far as the grandchildren were concerned, all they saw was a bird-like old woman who smoked like a chimney. Klaus remembers his mother Gret saying: 'A normal person would never love such a woman like Toni Wolff.' Even their grandfather made rude comments about her sometimes these days. But she retained a key role at the Psychological Club, and once the Jung Institute was founded in 1948 she was invited to give lectures there. So was Emma. Jung himself tried as much as possible to stay in the background. 'I remember when the Institute was just being organised and there was much nervous tension and criticism to be observed, the only person who was *not* a target was Mrs Jung,' recalled one student. 'I found myself wondering what could be wrong with Mrs Jung – until in the course of several months at the Institute, it became clear to me that everything was so right with Mrs Jung that nobody felt the need to find fault. She was admired, respected, beloved.' Another student recalled the difference in style between the two women: 'Emma Jung was still teaching at the Institute and contact with her in class or in her home was always warm and cordial. She was an ample woman, with an open and accepting attitude, very much her own person, who seemed to take the complexities and difficulties of life rather as a matter of course. She was always available for counsel and made me feel genuinely welcome.' Toni Wolff, in sharp contrast, appeared gaunt, haughty and forbidding. 'No smile crossed her face in class, in fact she betrayed no emotion of any kind. Questions were answered in clipped tones which made the questioner feel small, even stupid . . . The icy impression made us ponder how she could ever have been a "femme fatale" or "femme inspiratrice" to anybody, least of all C. G. Jung.'

People respected Toni Wolff's intellect but they had to know her very well before they discovered her inner warmth. And now even Carl, her life's partner, was turning away from her. In March 1953 she died, unexpectedly. She was sixty-five. She had been suffering from crippling arthritis for many years, but the week before she died she was still seeing patients and had attended the Psychological Club. The previous week she had gone to Seestrasse, and, quite unusually, stayed for the afternoon as well as lunch with both Emma and Carl. As the news spread of her death people turned up at the Jung house, not knowing where else to go. Emma received them courteously and gave them tea. A week later she attended the funeral service in the Peterskirche in Zürich. Carl stayed away. Some thought it heartless. Others thought he could not trust himself not to break down.

Later he carved a memorial stone with Chinese letters in her memory, and placed it in the garden at Seestrasse. 'Lotus. Nun. Mysterious', it read. But the stone gives it away. Whenever Carl was 'dissociated' he found a stone, held it, carved it, placed it.

'I shall always be grateful to Toni,' Emma said later with typical generosity, 'for doing for my husband what neither I nor anyone else could have done for him at a most critical time.'

Sabi Tauber, the daughter of a member of the Psychological Club, once asked Jung what was the happiest moment in his life. He answered that it had been when he was out sailing on the lake, after finishing *Answer to Job*, and he heard his father's voice saying: 'You have done the right thing, and I thank you for that . . .' Hearing this voice, he felt that 'he redeemed his father'. It was always his father, as Freud said.

On Valentine's Day 1953, Emma and Carl celebrated their golden wedding anniversary: fifty years of marriage. Surrounded by their children and grandchildren and a few close friends, they luncheoned, listened to some heartfelt speeches, and drank a toast to their achievement. Two years later it was Carl's eightieth birthday. The Kusnacht band came to play in the garden, and Emma and Carl danced a waltz on the terrace. But as it turned out, Emma was already ill with the cancer that would kill her a few months later.

Emma had had an operation that spring, 1955, but by November the metastasis had returned. Ruth Bailey, visiting from England, found Emma bedridden and Jung taking over the running of the house. On 22 November their family doctor telephoned with the test results. The family were gathered in the *Stube*, waiting, the women knitting or reading, the men smoking and walking about. Jung went into the hall to take the call: Emma only had days to live. When he came back into the room he was distraught, 'all tense and white – and not speaking'. Everyone was frightened to say anything. He told them the news and then he went upstairs to tell Emma. She took it calmly. Death appeared not to frighten her. During those last few days Andreas Jung remembers his mother telling him to go and see his beloved grandmother for the last time, to say goodbye. He found her reading a book about Eskimos. 'She was in bed, but we had a long and wonderful talk about Eskimos and how they got their green vegetables through eating sea-lions who had in turn eaten the ice moss.' Natural history had been Emma's first love all those years ago. Half an hour before she died her daughters Gret and Marianne were with her. Emma said, 'I'm going to die now, I'm going to say goodbye to you right now.' And so she did. Only Carl, her husband of fifty-two years, was with her when she died.

'I cannot continue the letter,' Jung wrote to a friend on 28 November. 'Yesterday I lost my wife after a serious illness lasting only five days. Please excuse me.' He had started the letter a week earlier but had not managed to finish it because of 'much anxiety'. Letters of condolence poured in but he couldn't answer them. 'Many thanks for your kind letters,' he finally wrote to Professor Boehler, a colleague from the ETH, on 14 December. 'The loss of my wife has taken it out of me, and at my age it is hard to recover.' To his old friend Erich Neumann he wrote describing the moment of 'illumination' he had experienced at Emma's bedside just before she died, which he was certain came from her:

Deepest thanks for your heartfelt letter. Let me in return express my deepest condolences on the loss of your mother. I am sorry I can only set down these dry words, but the shock I have experienced is so great that I can neither concentrate nor recover my power of speech. The quick and painless end – only five days between the final diagnosis and death – and this experience [the illumination] have been a great comfort to me. But the stillness and the audible silence about me, the empty air and the infinite distance, are hard to bear.

Michael and Frieda Fordham, an English couple who were part of the Jung circle, visited Carl some days after Emma's death to offer their condolences. 'We went to see him and really he could do nothing but weep,' remembered Frieda. 'He kept saying "She was a queen. She was a queen." We went to the funeral and went to see him at home afterwards. He was so broken up. He had become quite small, as if he had shrunk. It was quite an odd impression.' Ruth Bailey felt the same: 'he was sad for a long time when Emma went. It was touch and go whether he would ever come back, in a way.' The grandchildren were amazed to see their grandfather crying. 'He was terribly upset, he was really crying. I couldn't believe it,' said Jost Hoerni. Adrian agrees. 'I know that when she died he was extremely sad. Extremely. He had the feeling that a part of him had gone. He went to Bollingen, where he was really in grief. I think he was working something in stone.' No one knew what to do, and the family turned to Ruth for help. 'Aggi called after me and said "What do you do with him, Ruth, when he is like this? What can we do with him? I would not know what to do with him. What would you do with him?" I said "Oh, love him a bit. That is what he wants you to do, show him affection." Really, no one knew what to do.' None of them had ever seen him like this before.

'Mama's death has left a gap for me that cannot be filled,' Carl wrote to his daughter Marianne. 'The stone I am working on gives an inner stability with its hardness and permanence, and its meaning governs my thoughts.' The inscription was in Latin: 'My most beloved and faithful wife . . . she lived, she died, she suffered, she is missed . . .' Later he built a third and final tower at Bollingen, needing something even larger, more solid and permanent, to keep him steady.

The funeral service took place in the Protestant Reformed Church at Kusnacht on 30 November 1955. Emma Jung-Rauschenbach 30 March 1882–27 November 1955. The church was full, with people standing at the back, waiting. Then Carl walked in from a side door, followed by members of the Jung family, and took his place in the front row. The memorial oration was delivered by the Reverend Hans Schaer. 'Let every man be fully persuaded in his own mind,' he began, quoting St Paul to the Corinthians – a sure sign that this and what was to follow came direct from Jung, who quoted St Paul all his life, as did his father who was named after him. An interpretation of St Paul's words followed: 'All the threats, dangers and abysses to which man is exposed will not succeed in overpowering him but on the contrary will lead him further, and they will work together for his good. To be sure, this is only possible when man is guided by the strongest spiritual force, love.'

Then the Reverend Schaer came to Emma: 'a woman who has had a full life and has accomplished great tasks, leaves our midst; one who was a source of sustenance and strength to many. Rich gifts of mind and spirit had been entrusted to her. Life presented her with a rich fate and she carried it with great strength and in doing so was always true to herself.' It was a painful loss even to those who paid only occasional visits to the beautiful house on Seestrasse, where her living presence in the home was like a gift to all those who came there. But how much greater the loss of those who were near to her. It was a fulfilled life, and she was permitted to carry out fine and important tasks. She saw something good in everything and she was able to bestow good because she was persuaded and convinced in her own mind. Hers had been an idyllic childhood till tragedy struck. 'When she was only twelve years old her father became blind and this saddened her childhood. For the man who had been so active till then could not reconcile himself to his fate. His suffering caused many a bitter hardship and heartache to his family. She helped her mother carry the burden which was really beyond the strength of her years,' said the Reverend Schaer. Everything that was careless or artificial and theoretical was foreign to her. She had great kindness and modesty. She was always able to hold her own as an independent, at the side of her husband. 'With her sense of humour that saw through everything that was inflated or one-sided, she resolved difficulties and tensions and helped to overcome obstacles. She was the nurturing soil in which his [Jung's] creativity took root and from which he drew essential strength. A light emanated from her ... She was able to carry and endure life's burdens and above all she knew how to recognise and safeguard the secrets of others.' Meaning, surely, the secrets of her husband Carl.

If there was any doubt that the text had been written by Jung, it was banished now. No one in the church could be in any doubt that Carl Jung had finally made public his deep debt to his wife.

That afternoon the Jung family opened Seestrasse to the mourners. Carl did not come downstairs. He stayed up in the library, surrounded by his own thoughts.

'I saw her in a dream which was like a vision,' he wrote:

She stood at some distance from me, looking at me squarely. She was in her prime, perhaps about thirty, and wearing the dress which had been made for her many years before by my cousin the medium. Her expression was neither joyful nor sad, but, rather, objectively wise and understanding, without the slightest emotional reaction, as though she were beyond the mists of affects.

I knew that it was not she, but a portrait she had made or commissioned for me. It contained the beginning of our relationship, the events of fifty-three years of marriage, and the end of her life also.

Emma and Carl. Carl and Emma. 'Face to face with such wholeness one remains speechless, for it can scarcely be comprehended.'

Notes

Abbreviations used in the Notes

CW: C. G. Jung, *The Collected Works of C. G. Jung*, 20 vols, Bollingen Series, Princeton, NJ: Princeton University Press, 1960–90 (see Bibliography for individual volumes)

Freud/Jung Letters: Sigmund Freud and C. G. Jung, *The Freud/ Jung Letters: The Correspondence between Sigmund Freud and C. G. Jung*, ed. William McGuire, tr. Ralph Manheim and R. F. C. Hull, Princeton, NJ: Princeton University Press, 1974

Interview with G. F. Nameche: the interviews conducted 1969–70 by G. F. Nameche are held at the C. G. Jung Biographical Archive, Countway Library of Medicine, Harvard Medical School

MDR: C. G. Jung and Aniela Jaffé, *Memories, Dreams, Reflections*, New York: Pantheon, 1962

Chapter 1: A Visit to Vienna

2 'Psychoanalysis': the term went through many variations during the early years.

3 3,000 Swiss francs: *MDR*, p. 101.

3 a very large sum: Swiss money value in Rolf Mösli, *Eugen Bleuler: Pionier der Psychiatrie*, Zürich: Römerhof Verlag, 2012, p. 116.

3 In the event of a divorce: with thanks to Dr Elizabeth Schlumpf in Zürich for legal details taken from Peter Tuor et

al., *Das Schweizerische Zivil-Gesetzbuch*, p. 139: 'End of Marriage and Divorce, 1905: in case of divorce, each partner takes out of the marriage what they originally brought in. Any subsequent incoming extras to be divided two-thirds to the husband, one-third to the wife.'

4   'I shall be in Vienna': *Freud/Jung Letters*, p. 24.

4   that March morning: for the first meeting see Ludwig Binswanger, *Sigmund Freud: Reminiscences of a Friendship*, tr. Norbert Guterman, New York: Grune & Stratton, 1957, p. 10; also Eva Weissweiler, *Die Freuds: Biographie einer Familie*, Cologne: Kiepenhauer & Witsch, 2006, p. 139, and C. G. Jung, *Letters of C. G. Jung*, ed. Gerhard Adler and Aniela Jaffé, tr. R. F. C. Hull, Princeton, NJ: Princeton University Press, 1973–75, Vol. 2, p. 36.

5   'Sundays I am free': *Freud/Jung Letters*, p. 23.

6   Freud's self-deprecating Jewish humour: Deirdre Bair, *Jung: A Biography*, Boston, MA: Little, Brown, 2003, p. 119.

7   On special occasions: for the Sunday lunch see Martin Freud, *Glory Reflected: Sigmund Freud – Man and Father*, London: Angus & Robertson, 1957, p. 108.

9   'In my experience': *Introduction to Jungian Psychology: Notes of the Seminar on Analytical Psychology Given in 1925*, ed. and Introduction by William McGuire, tr. R F. C. Hull, new Introduction and updates by Sonu Shamdasani, Princeton: Princeton University Press, 1989, p. 20, and *MDR*, p. 146.

10 'Our Aryan comrades': John Kerr, *A Most Dangerous Method: The Story of Jung, Freud, and Sabina Spielrein*, New York: Knopf, 1993, p. 384.

11 'like a small boy': Martin Freud, *Glory Reflected*, p. 109.

12 the Grand: for the Grand Hotel see Karl Baedeker, *Switzerland: And the Adjacent Portions of Italy, Savoy, and Tyrol*, Leipzig: Karl Baedeker, 1905.

13 female hysterics with sexual dysfunction: Maines, Rachel P., *The Technology of Orgasm: 'Hysteria', the Vibrator, and*

Women's Sexual Satisfaction, Baltimore, MD: Johns Hopkins University Press, 1999.

14 'In the following pages': Sigmund Freud, *The Interpretation of Dreams*.

14 'It seems to me': *Freud/Jung Letters*, pp. 4, 14, and as discussed in George Makari, *Revolution in Mind: The Creation of Psychoanalysis*, London: Duckworth, 2008, p. 203.

14 that particular Wednesday: for the Wednesday meeting see Binswanger, *Sigmund Freud*, p. 11.

15 'I felt so foreign': Kurt Eissler interview, 1953, Sigmund Freud Collection, Manuscript Division, Library of Congress, Washington DC, in Bair, *Jung*, p. 118.

15 'Now you've seen': Binswanger, *Sigmund Freud*, p. 11.

15 When Freud joked: for the Jewish joke see Bair, *Jung*, p. 119.

17 Worse was to come: for Carl's flirtation see *Freud/Jung Letters*, pp. 229, 24 and 25.

Chapter 2: Two Childhoods

19 Emma recalled her childhood: the description of Emma's childhood is based on interviews with the Jung grandchildren (see Chapter 15) and Beatrice Homberger; *Jugend-Erinnerungen einer Grossmutter* by Gertrud Henne, privately printed 1960, quoted in Andreas Jung, Regula Michel, Arthur Ruegg, Judith Rohrer and Daniel Ganz, *The House of C. G. Jung: The History and Restoration of the Residence of Emma and Carl Gustav Jung-Rauschenbach*, Wilmette, IL: Chiron Publications, 2008; and the family memoirs of Helene Hoerni-Jung, with thanks to Carl and Emma's grandson Adrian Baumann.

20 'If only you'd remained': *Schaffhauser Biographen*, p. 21, with thanks to the Stadtarchiv Schaffhausen.

23 'the young, very pretty': *MDR*, p. 23.

24 Carl first clapped eyes: for their first meeting see Alan C. Elms, *Uncovering Lives: The Uneasy Alliance of Biography*

and Psychology, Oxford: Oxford University Press, 1994, p. 61.

24 the first correspondence between them: with thanks to Andreas Jung, who checked this in the family archive.

25 Franz confirmed years later: the interview with Franz is in Linda Donn, *Freud and Jung: Years of Friendship, Years of Loss*, New York: Charles Scribner's Sons, 1988, pp. 61 and 62.

26 'My situation is mirrored': Aniela Jaffé, *C. G. Jung: Word and Image*, Bollingen Series XCVII, Vol. 2, Princeton, NJ: Princeton University Press, 1979, p. 27.

27 'My stone,' he called it: *MDR*, p. 33.

29 'Dim intimations of trouble': *MDR*, p. 22.

29 'I had never come across': William McGuire and R. F. C. Hull (eds), *C. G. Jung Speaking: Interviews and Encounters*, London: Picador, 1978, p. 25.

30 The first of these was a dream: *MDR*, p. 25.

31 'a little manikin': *MDR*, p. 34.

33 'as timid and craven': *MDR*, p. 39.

33 'disagreeable, rather uncanny feeling': Bair, *Jung*, p. 30.

33 As he described it: *MDR*, p. 42.

35 'The world is beautiful': *MDR*, p. 47.

36 'My entire youth': *MDR*, p. 52.

36 on the rock stood a castle: *MDR*, p. 86.

39 'I was seized with': *MDR*, p. 64.

39 the tragedy of his youth: Jaffé, *C. G. Jung*, p. 20.

39 'I was sure': *MDR*, p. 58.

40 'The feeling I associated': *MDR*, p. 23.

40 'The following days': *MDR*, p. 100.

40 a stipend to help fund: *MDR*, p. 92.

41 'intellectually dominating': McGuire and Hull (eds), *C. G. Jung Speaking*, p. 29.

41 the nickname Walze: *Walze* has often been translated as the 'barrel' or 'log', but 'Steam-Roller' is more accurate, and describes Carl's personality as well as his appearance.

41 'He was appalled': McGuire and Hull (eds), *C. G. Jung Speaking*, p. 30.

42 'Your last two letters': *Freud/Jung Letters*, p. 94.

Chapter 3: A Secret Betrothal

45 'My heart suddenly began to pound': *MDR*, p. 111.

46 'The ideal wife': Albert Tanner, *Arbeitsame Patrioten – wohlanständige Damen: Bürgertum und Bürgerlichkeit in der Schweiz, 1830–1914*, Zürich: Orell Füssli Verlag, 1995, p. 209, author's translation.

47 'The happiness and lasting power': Rosa Dahinden-Pfyl, *Die Kunst mit Männern Glücklich zu Sein*, p. 219, author's translation.

48 'He didn't think much of': McGuire and Hull (eds), *C. G. Jung Speaking*, p. 28.

49 a difficult, sarcastic man: see Jean Rauschenbach's obituary in *Tagesblatt*, with thanks to Stadtarchiv Schaffhausen.

49 Herr Rauschenbach had syphilis: this sad story is based on author interviews with the Jung family. A dramatic illustration is Rembrandt's portrait of the artist Gerard de Lairesse with his grossly deformed face, who, like Emma's father, eventually went blind.

51 in the medical pamphlets: see Claude Quétel, *History of Syphilis*, tr. Judith Braddock and Brian Pike, Cambridge: Polity Press, 1990.

52 'psychologically abnormal': Bair, *Jung*, p. 59.

52 'the utmost concentration': Jung, *Introduction to Jungian Psychology*, and Bair, *Jung*, p. 59.

53 'how the human mind reacted': *MDR*, p. 114.

53 She worried: author interviews with the Jung family.

54 'Mein liebster Schatz!': with thanks to Andreas Jung for confirmation of this form of address.

54 The Burghölzli at the turn of the twentieth century: with thanks to Herr Rolf Mösli, first archivist at the Burghölzli

and author of an excellent book on Bleuler (*Eugen Bleuler: Pionier der Psychiatrie*, Zürich: Römerhof Verlag, 2012), who was most generous with his time.

54 'In the medical world at the time': *MDR*, p. 111.

57 there were seventy *Wärters*: Mösli, *Eugen Bleuler*, p. 138. Mösli's book is also excellent on daily life at the Burghölzli.

59 'It was as though two rivers': *MDR*, p. 112.

60 the woman in Carl's section: for the depressive woman see *MDR*, p. 116.

62 'a girl with poor inheritance': C. G. Jung, *CW1: Psychiatric Studies*, ed. Herbert Read, Michael Fordham and Gerard Adler, tr. R. F. C. Hull, Princeton, N.J.: Princeton University Press, 1970, p. 17.

62 'vaulting ambition': *Freud/Jung Letters*, p. 78.

62 'the progressive elucidation': Jung, *CW1* (author's own translation as it appears only in the German version).

63 'a submission to the vow': *MDR*, p. 114. Jung describes this as his feelings at the start of his time at the Burghölzli but it more accurately describes how he felt later, once he had got used to the work and the routine of the institution. At first, by his own account, he was completely overwhelmed.

64 Swiss army military service: with thanks to Othmar Zeltner, who began his military service in St Gallen in 1952.

65 'in the style of the eighteenth century': Bair, *Jung*, p. 80, and Andreas Jung confirms that Carl had an English teacher in London. This is probably how Carl came to be in Oxford.

66 'Seine landscape with clouds': Jaffé, *C. G. Jung*, p. 42.

Chapter 4: A Rich Marriage

67 a church wedding in the Steigkirche: the wedding details based on author interviews with Jung grandchildren (see Chapter 15), and with thanks to David Syffer at the IWC archive, Schaffhausen.

67 luckily there was no snow: the wedding-day weather is thanks to Nicola Behrens at the Stadtarchiv Zürich.

67 The wedding banquet: Gerhard Wehr, *An Illustrated Biography of C. G. Jung*, tr. Michael Kohn, Boston: Shambhala Publications, 1989, p. 25.

69 'The further away from church': *MDR*, p. 81.

69 setting off on their honeymoon: With thanks to Andreas Jung for the details of their itinerary.

69 Continental Express to Paris: thanks to SSB Historical Archives in Bern for answering queries about the comforts of first-class travel and providing timetables and illustrations.

70 a quarrel about money: Vincent Brome, *Jung: Man and Myth*, London: Macmillan, 1978, p. 83.

70 Swiss law gave the husband: with thanks to Dr Elizabeth Schlumpf in Zürich for legal details from Dr P. Tuor, *Das Schweizerische Zivil Gesetzbuch*.

70 Emma had seen Carl's temper: for Carl's violent rages see *MDR*, p. 54, and Jaffé, *C. G. Jung*, p. 132.

71 'an adventure before marriage': Bair, *Jung*, p. 70, and that Carl was an 'absolutely reliable husband' for the first seven years or so of marriage.

71 a rented apartment in Zürich: for Zurich in 1903 see Kurt Guggenheim, *Alles in Allem, Zürich: Artemis Verlag, 1954*; Walter Baumann, *Zürich: La Belle Époque*, Zürich: Orell Füssli Verlag, 1973, and Zürcher Schlagzeilen, Zürich: Orell Füssli Verlag, 1981; Hanspeter Danuser, *Das Trambuch: 100 Jahre Züri-Tram*, Zürich: Verlag Neue Zürcher Zeitung, 1982.

75 entry costing 60 rappen: There are 100 rappen to the Swiss franc.

77 Marguerite's fiancé, Ernst Homberger: author interview with Ernst and Margeurite's granddaughter Beatrice Homberger.

79 'I'm sitting here in the Burghölzli': C.G. Jung letter to Andreas Vischer in Angela Graf-Nold, 'The Zürich School of Psychiatry in Theory and Practice: Sabina Spielrein's

Treatment at the Burghölzli Clinic in Zürich', *Journal of Analytical Psychology*, Vol. 46, Issue 1, January 2001, p. 83.

80 'As usual you have hit the nail on the head': *Freud/Jung Letters*, p. 78.

83 'The years at Burghölzli were my apprenticeship': *MDR*, p. 116.

83 Hedwig, who was a remarkable woman: for Hedwig Bleuler see 'Erinnerungen an Hedwig Bleuler-Waser', *Zentralblatt* 20, April 1940; Mösli, *Eugen Bleuler*; Hedwig Bleuler-Waser, 'Aus meinem Leben', *Schweizer Frauen der Tat* 3, 1929.

86 Other researchers used whistles: C. G. Jung, *Studies in Word Association*, in *CW2: Experimental Researches*, ed. and tr. Gerhard Adler and R. F. C. Hull, Princeton, NJ: Princeton University Press, 1973, pp. 5, 228.

86 diagnoses and classification of: Jung, *Studies in Word Association*, p. 6.

86 the 'one-fifth-second stop watch': with thanks to David Syffer at IWC Schaffhausen

87 'If we ask patients directly': Jung, *Studies in Word Association*, p. 298.

88 Emma appears as 'Subject no. I': Jung, *Studies in Word Association*, p. 240.

88 'The subject is pregnant': for Emma's pregnancy complex see Jung, *Studies in Word Association*, p. 243.

Chapter 5: Tricky Times

92 Emma breastfed hers herself: see *Freud/Jung Letters*, p. 188, 'My wife is, of course, nursing the child herself, a pleasure for both of them.'

93 more conventional roles: for a woman's place in the society of the time see Tanner, *Arbeitsame Patrioten*, pp. 226 and 227.

94 Martha became pregnant six times in ten years: Katja Behling, *Martha Freud: A Biography*, Cambridge: Polity Press, 2005, p. 113.

94 their marriage becoming 'amortised': *Freud/Jung Letters*, p. 456.

95 Ernst Homberger, who had married: with thanks to Beatrice Homberger.

95 stated the obituary: in the *Schaffhausen Tagesblatt*, 3 March 1905, with thanks to the Stadtarchiv Schaffhausen.

97 'My aim was to show': *MDR*, pp. 112 and 118.

97 no shortage of stories: *MDR*, pp. 123, 124.

98 'He kept his students spellbound': Binswanger, *Sigmund Freud*, p. 1.

98 *Pelzmäntel* – fur-coat – brigade: Bair, *Jung*, pp. 98 and 109.

99 charming, fuzzy snapshots: the photographs are in the Jung family archive, with thanks to Andreas Jung.

99 Emma loved her children: based on author interviews with the Jung family.

100 two female inmates: Jung, *Studies in Word Association*, pp. 381 and 382.

101 his cousin Helly: Jung, *Introduction to Jungian Psychology*, p. 6

101 the night of 17 August 1904: Graf-Nold, 'The Zürich School of Psychiatry in Theory and Practice', p. 73.

101 Baur-en-Ville: there were two hotels: Baur-au-Lac and Baur-en-Ville.

102 The young woman was Sabina Spielrein: the Spielrein detail and her history are based on Bernard Minder, 'Sabina Spielrein: Jung's Patient at the Burghölzli', *Journal of Analytical Psychology*, Vol. 46, Issue 1, January 2001, and Graf-Nold, 'The Zürich School of Psychiatry'.

105 *Punch, the Illustrated London News*: from a family memoir, with thanks to Adrian Baumann.

105 The *Neue Zürcher Zeitung* felt: the news reports are from Baumann, *Zürcher Schlagzeilen*, p. 17.

107 leaving it at that: Graf-Nold, 'The Zürich School of Psychiatry', p. 93. Graf-Nold pointed out in an interview with the author how strange it was that Jung made no

further comment about the clear indications of sexual abuse.

111 Jung's report, dated 25 September 1905: Minder, 'Sabina Spielrein', p. 68.

Chapter 6: Dreams and Tests

113 'I saw horses': Jung *CW3*, p. 57.

119 'You have put your finger': *Freud/Jung Letters*, p. 14.

120 'I broke it off': *Freud/Jung Letters*, p. 17.

121 Three crosses meant danger: *Freud/Jung Letters*, p. 19.

122 Bleuler's 'active community': Graf-Nold, 'The Zürich School of Psychiatry', p. 90.

122 'In the hospital the spirit of Freud': Abraham Brill, *Lectures on Psychoanalytic Psychiatry*, New York: Knopf, 1946, quoted in Kerr, *A Most Dangerous Method*, p. 172.

123 'This Freud cult': Mikkel Borch-Jacobsen and Sonu Shamdasani, *The Freud Files: An Inquiry into the History of Psychoanalysis, Cambridge: Cambridge University Press, 2012*, p. 69.

123 'a German-American woman': *Freud/Jung Letters*, p. 71.

123 'My wife, who knows a thing': *Freud/Jung Letters*, p. 72.

124 'But don't you think': *Freud/Jung Letters*, p. 4.

124 'I had grown up in the country': *MDR*, p. 161.

125 'As I have indicated before': *Freud/Jung Letters*, p. 229.

125 'the trauma [was] wished for': Angela Graf-Nold, '100 Jahre Peinlichkeit – und (k)ein Ende?: Karl Abrahams frühe psychoanalytische Veröffentlichungen und die "sexuelle Frage"', *Jahrbuch der Psychoanalyse 52 (2006): 93–138*, pp. 52, 127.

125 'You will doubtless have drawn': Mc Guire, pp. 25 and 144.

126 'Often I want to give up': *Freud/Jung Letters*, pp. 75 and 89.

126 'refusing to obey the chairman's repeated signal': Borch-Jacobsen and Shamdasani, *The Freud Files*, p. 74.

127 'I have a sin to confess': *Freud/Jung Letters*, p. 115.

127 'At the moment I am treating': *Freud/Jung Letters*, p. 108.

128 she threatened Carl with divorce: Susi Trüb, interview with G. F. Nameche, and author interviews with the Jung family.

129 caught out by a transgression: author interviews with the Jung family.

129 'All sorts of things': *Freud/Jung Letters*, p. 117.

130 'I will begin with an experiment': Ludwig Binswanger, 'On the Psychogalvanic Phenomenon in Association Experiments', in Jung, *Studies in Word Association*, p. 457.

131 'My veneration for you': *Freud/Jung Letters*, p. 95.

132 'in a state of active opposition': Ludwig Binswanger, 'On the Psychogalvanic Phenomenon in Association Experiments', in Jung, *Studies in Word Association*, p. 497.

133 'He is also reserved': A better translation would be 'defensive', not 'reserved'.

Chapter 7: A Home of Their Own

135 The rumours flying about the place: *Freud/Jung Letters*, p. 207, and see Chapter 8.

135 They purchased a plot of land: Barbara Hannah, *Jung: His Life and Work*, London: Michael Joseph, 1976, p. 93: They found it 'more or less by chance' when out for a Sunday walk. They saw a land for sale sign. 'Her eyes still shone when she described the joy and excitement this aroused in them both.'

136 'No. 1 wanted to free himself': *MDR*, p. 86.

137 Within a month of buying the plot: The information about the house comes from Andreas Jung et al., *The House of C. G. Jung*. With thanks to Andreas Jung, who lives in the Jung house and showed me round on more than one occasion, and showed me the original sketches.

139 'Much of it struck us': Jung et al., *The House of C. G. Jung*, p. 20.

142 'We are in complete agreement': Jung et al., *The House of C. G. Jung*, p. 40.

145 they preferred to stay at the Hotel Regina: *Freud/Jung Letters*, p. 213.

145 'to begin with, Jung': Jones, *The Life and Work of Sigmund Freud*, Vol. 2, pp. 254 and 258.

146 'My selfish purpose': *Freud/Jung Letters*, p. 168.

146 Freud came to visit Jung: Jones, *The Life and Work of Sigmund Freud*, Vol. 2, p. 264.

147 'a surprise package of books': *Freud/Jung Letters*, p. 173.

147 Emma stayed with her mother at Ölberg: details of life at Ölberg based on author interviews with Carl and Emma's granddaughter Brigitte Merk and Beatrice Homberger, and family memoir, with thanks to Adrian Baumann.

148 the hot weather could turn heavy: with thanks to Nicola Behrens at Stadtarchiv Zürich.

148 Emma put on her jewels: author interview with Adrian Baumann.

149 Carl only visited at the weekends: for Carl's work and holiday see Graf-Nold, 'The Zürich School of Psychiatry in Theory and Practice', p. 175: Writing to Sabina Spielrein: 'First I am going for a week to Toggenburg and will be walking with Riklin, then a further six days to Schaffhausen to my wife and child(ren).'

149 'fleeing into the inaccessible solitude': *Freud/Jung Letters*, p. 171.

149 'Babette S': *MDR*, p. 126.

153 the boy, Manfred: Susi Trüb, interview with G. F. Nameche, and author interview with Rolf Mösli.

153 the new life: see McGuire's note in *Freud/Jung Letters*, p. 215. Carl continued as *Privatdozent* lecturer at Zürich University till April 1914.

153 'This last miserable week': *Freud/Jung Letters*, p. 224.

154 the majority of toilets in Switzerland: many were still that basic into the 1950s, including in the author's Swiss grandparents' house.

155 leading to the library: For Jung's collection of books see Sonu Shamdasani, *C. G. Jung: A Biography in Books*, New York: W. W. Norton & Company, 2012.

Chapter 8: A Vile Scandal

158 Franz Karl Jung was born: *Freud/Jung Letters*, pp. 183, 184, 188.

159 four-year-old Agathli's reaction: *Freud/Jung Letters*, pp. 199, 208 and 212.

160 'I regret so much': Graf-Nold, 'The Zürich School of Psychiatry in Theory and Practice', p. 177.

161 his treatment of Otto Gross: Makari, *Revolution in Mind*, p. 231, Otto Gros 'a sexual revolutionary'. Analysis sessions lasted up to twelve hours when the tables were sometimes turned and the patient became the doctor; and *Freud/Jung Letters*, pp. 157, 156, 153, 155.

162 sexual repression was 'a very important': *Freud/Jung Letters*, p. 90.

163 'My dear Fräulein Spielrein': Graf-Nold, 'The Zürich School of Psychiatry in Theory and Practice', p. 173.

163 a suitcase belonging to Sabina Spielrein: Graf-Nold, 'The Zürich School of Psychiatry in Theory and Practice', 155.

164 'It is my misfortune': Graf-Nold, 'The Zürich School of Psychiatry in Theory and Practice', 177.

165 she wrote an anonymous letter: Aldo Carotenuto, *A Secret Symmetry: Sabina Spielrein Between Jung and Freud*, New York: Pantheon Books, 1982, p. 93. I disagree with Bair, *Jung*, p. 191, who writes that Carl acquiesced to Emma's every demand from that time onward: from his Burghölzli resignation to the buying of Kusnacht house. Carl wanted both these himself. Nor was it 'divorce proceedings' – threats, certainly, and no evidence of the pregnancy being 'deliberate'. My interviews with the the Jung family support this.

165 'under terrific strain': *Freud/Jung Letters*, p. 207.

166 'I have always told your daughter': for whether sexual relationship was sexual see Carotenuto, *A Secret Symmetry*, p. 95.

167 'To be slandered': *Freud/Jung Letters*, p. 210.

168 'I've never really had a mistress': *Freud/Jung Letters*, p. 212.

168 'Hurrah for your new house!': *Freud/Jung Letters*, p. 226.

168 'She was, of course, systematically planning my seduction': *Freud/Jung Letters*, p. 228.

169 'Such experiences, though painful': *Freud/Jung Letters*, p. 230.

170 'Although not succumbing': *Freud/Jung Letters*, p. 236.

171 'risk his authority!': *MDR*, p. 154.

171 'Father was very disappointed': Donn, *Freud and Jung*, p. 98.

172 '*Liebste Frau!*': Jaffé, *C. G. Jung*, p. 47.

174 They had taken the 'elevated': for the journey to Worcester see *MDR*, p. 336.

177 Joseph Medill McCormick: William McGuire, 'Firm Affinities: Jung's Relations with Britain and the United States', *Journal of Analytical Psychology*, 40.3, July 1995, pp. 307, 305.

178 '*Mein liebster Schatz*': Jaffé, *C. G. Jung*, p. 49.

180 'Two more days before departure!': *MDR*, p. 338.

Chapter 9: Emma Moves Ahead

181 'The sofa cushion crawled about': *MDR*, p. 339.

183 'In my family all is well': *Freud/Jung Letters*, p. 252.

183 Unterwasser, a secluded Alpine village: *Freud/Jung Letters*, p. 284.

183 She had a fairly objective view: Sigmund Freud and Ernest Jones, *The Complete Correspondence of Sigmund Freud and Ernest Jones, 1908–1939*, ed. R. Andrew Paskauskas, Cambridge, MA: Harvard University Press, 1993, Jones to Freud, 18 September 1912, p. 155.

184 There was no lack of women: for Carl's women see Carotenuto, *A Secret Symmetry*, p. 17, and Kerr, *A Most Dangerous Method*, p. 299.

184 on a good day: for Spielrein's ups and downs see Carotenuto, *A Secret Symmetry*, pp. 12–17.

185 the secret diary he had kept: Carotenuto, *A Secret Symmetry*, p. 12.

186 'To be fruitful': C. G. Jung, *CW5: Symbols of Transformation*, ed. and tr. Gerhard Adler and R. F.C. Hull, Princeton, NJ: *Princeton University Press, 1977*, and Kerr, *A Most Dangerous Method*, p. 327.

187 that night Jung had a dream: *MDR*, p. 284 and *Freud/Jung Letters*, p. 359, fn1.

188 '*Sehr Geehrter Herr Professor!*': for Emma's letters see *Freud/Jung Letters*, pp. 301 and 303.

189 'Now don't get cross': *Freud/Jung Letters*, p. 302.

189 'Bleuler is not coming': *Freud/Jung Letters*, p. 304. The notes add that the International Psychoanalytic Association was founded at the Congress, with Jung elected president.

189 the pre-eminence of the Freudian school: See Borch-Jacobsen and Shamdasani, *The Freud Files*, for an excellent analysis.

189 'Most of you are Jews': Kerr, *A Most Dangerous Method*, p. 287.

190 'gadding about like mad': *Freud/Jung Letters*, p. 341.

191 a heatwave hit the Continent: Guggenheim, *Alles in Allem*, Vol. 1, p. 229.

192 'Charming, clever and ambitious': *Freud/Jung Letters*, p. 436.

193 a pedagogic congress in Brussels ... mountain tour: *Freud/Jung Letters*, p. 439.

194 'This time the feminine element': *Freud/Jung Letters*, p. 440

194 'Dear friend,' Freud replied: *Freud/Jung Letters*, p. 441.

192 'There were of course seminars': Ernest Jones quoted by McGuire in note to *Freud/Jung Letters*, p. 443.

197 'Dear Professor Freud': For Emma's letters to Freud see *Freud/Jung Letters*, pp. 452–67.

201 '"he will grow, but I must dwindle"': 'I must dwindle' suggests Emma has heard Carl's views on this too, not only Fräulein Spielrein.

205 'tendency to autoeroticism': the term 'autoeroticism' appears to have been used in various ways at the time, and it is impossible to know what exactly Emma meant by it, but most likely it was just a general concern with oneself.

Chapter 10: A Difficult Year

206 a difficult year for Zürich: for Zürich in 1912 see Guggenheim, *Alles in Allem*, Vol. 1, p. 259.

206 the automobile factory at Schlieren: for the automobile statistics see Tanner, *Arbeitsame Patrioten*, p. 399.

208 'mighty rumblings in Zürich': *Freud/Jung Letters*, p. 484.

210 the state visit by Kaiser Wilhelm II: Guggenheim, *Alles in Allem*, Vol. 1, p. 265.

211 his first *Kriegsrat*: Clay, p. 297.

213 'It is of course ironical': *MDR*, p. 181.

213 'When working on my book': *MDR*, p. 162.

214 Jung was 'a born leader': *The Correspondence of Sigmund Freud and Sándor Ferenczi*, Volume 1, *1908–1914*, ed. Eva Brabant, Ernst Falzeder and Patrizia Giampieri-Deutsch, tr. Peter T. Hoffer, Cambridge, MA: Harvard University Press, 1993, p. 434. Note: it is not clear whether Freud is quoting his conversation with Jung at Munich. But either way, this was his opinion.

214 I have a wife and [five] children': In fact, four. Helene was born in 1914.

214 'The unconscious contents': *MDR*, p. 181

215 'I read your essay': *Freud/Jung Letters*, p. 331, and fn3.

216 'It seemed to me': Jung, *Introduction to Jungian Psychology*, and C. G. Jung, *The Red Book: Liber Novus*, ed. Sonu

Shamdasani, tr. M. Kyburz, J. Peck and S. Shamdasani, New York: W. W. Norton & Company, 2009, p. 12.

217 'I make horoscope calculations': *Freud/Jung Letters*, p. 427.

217 the theory of the death wish: Kerr, *A Most Dangerous Method*, p. 403.

217 'I have tried': *Freud/Jung Letters*, p. 288.

217 deep depression: Hannah, *Jung*, p. 99.

217 'It took me a long time': Jung, *Introduction to Jungian Psychology*, p. 28.

217 'While working on the book': *MDR*, p. 162.

218 'So perhaps it is all to the good': *Freud/Jung Letters*, p. 456.

218 'I had to tell myself this': *Freud/Jung Letters*, p. 232.

219 the 'enormous turd': *MDR*, p. 50.

219 he would tell Emma: see the funeral oration in the Epilogue.

219 another of his trips to America: for Jung's 1912 trip to America see McGuire and Hull (eds), *C. G. Jung Speaking*, Introduction and pp. 29 and 39.

223 'The journey from cloud-cuckoo-land': see McGuire's Introduction to *Freud/Jung Letters*.

224 his little tricks: *Freud/Jung Letters*, p. 535.

224 'As regards Jung': Freud and Jones, *The Complete Correspondence*, p. 186.

224 'behaving like a florid fool', *The Correspondence of Sigmund Freud and Sándor Ferenczi*, p. 353.

225 another significant dream: *MDR*, p. 166.

225 a memory from his eleventh year: *MDR*, p. 168.

226 'Father would be down there': For Franz's quotes see Donn, *Freud and Jung*, pp. 160, 172 and 174.

227 my father could do no work: clearly Franz was wrong about that.

228 'It was most essential': *MDR*, pp. 181, 182.

Chapter 11: *Ménage à Trois*

229 'staged a number of jealous scenes': *Freud/Jung Letters*, p. 289.

230 'the devil can use': *Freud/Jung Letters*, p. 207.

230 'I really cannot remember': *Freud/Jung Letters*, p. 351.

230 many men in his circle kept mistresses: Kerr, p. 379.

230 his unmarried sister-in-law, Minna: this theory is discussed in Kerr, *A Most Dangerous Method*, p. 135.

230 'I stand for an infinitely freer sexual life': Lisa Appignanesi and John Forrester, *Freud's Women*, London: Weidenfeld and Nicolson, 1992, p. 51.

231 'My Indian summer': *Freud/Jung Letters*, p. 292.

231 Maria Moltzer: *The Correspondence of Sigmund Freud and Sándor Ferenczi*, p. 446, with thanks to Sonu Shamdasani for confirmation. The full quote from Freud to Ferenczi: 'The master who analysed him can only have been Fräulein Moltzer, and he is so foolish as to be proud of this work of a woman with whom he is having an affair.'

232 Emma threatened divorce three times: Susi Trüb, interview with G. F. Nameche, and Ronald Hayman, *A Life of Jung*, London: Bloomsbury, 1999, p. 126, and author interviews with the Jung family.

232 'Except for moments of infatuation': *Freud/Jung Letters*, p. 212.

232 'A new discovery of mine': *Freud/Jung Letters*, p. 440.

234 'A woman who does not have': Bair, *Jung*, p. 197.

235 'very near psychosis': Carl Meier, interview with G. F. Nameche.

235 Jung used to say: Donn, *Freud and Jung*, p. 179.

235 'I was furious': Kerr, *A Most Dangerous Method*, p. 373 and *Freud/Jung Letters*, p. 465.

235 'With us everything is peaceful': *Freud/Jung Letters*, p. 484.

235 'never totally in life': Susi Trüb, interview with G. F. Nameche.

235 'Toni was all spirit': Donn, *Freud and Jung*, p. 178; for further descriptions of Toni Wolff see Maggy Anthony, *The Valkyries: The Women Around Jung*, Longmead: Element Books, 1990, p. 33.

236 'In her presence': Tina Keller, interview with G. F. Nameche.

236 'After the parting of ways': *MDR*, p. 165.

237 feared he had a psychic disturbance: for Jung's illness see Brome, *Jung*, pp. 162, 301. D. W. Winnicott, reviewing Jung's *MDR* in the *International Journal of Psychoanalysis*, Apr–Jul 1964, describes him as a 'recovered case of infantile psychosis', giving us 'a picture of childhood schizophrenia' which settled down into a 'splitting of the personality'. Also with thanks to Sonu Shamdasani for pointing me to Henri Ellenberger's concept of Jung's 'creative illness' in *The Discovery of the Unconscious*.

237 laughing 'down to his shoes': Donn, *Freud and Jung*, p. 90.

237 talking away to Philemon: *MDR*, p. 176.

237 'Then I came to this': Jung, *Introduction to Jungian Psychology*, p. 45, with thanks to Sonu Shamdasani for confirmation that it was Moltzer's voice.

238 this time in Munich: Hayman, *A Life of Jung*, p. 169.

238 'considers Jung to be mentally disordered': Freud and Jones, *The Complete Correspondence*, p. 237, 11 November 1913. Freud writes back to Jones on 17 November: 'I have been very much amused by your good saying about Jung, but we must not forget, this is our only case of success in the campaign against him, and we are mostly dependent on him for helping us by his foolish ways' – revealing how there was now a concerted campaign to discredit Jung.

238 'When I had the vision': *MDR*, p. 169, and C. G. Jung, *Liber Novus*, Chapter 1.

239 Jung's visionary *Liber Novus*: see Shamdasani's Introduction to C. G. Jung, *Liber Novus*, pp. 48 and 127 (fn34) on the start of the writing – and a fine introduction to the whole.

240 it lay undisturbed in his *Cabinet*: in 1983 it was placed in a safe box and a year later five copies were made. Finally, in 2009, a facsimile edition was made at the behest of the Jung family, a superb achievement of labour and love by Sonu Shamdasani and the translators, and the publishers W. W. Norton and Company, handsomely supported by the Philemon Foundation.

122 And Emma: Ulrich Hoerni (Carl and Emma's grandson), in the Preface to C. G. Jung, *Liber Novus*.

240 'It was as though the ground': *MDR*, p. 172.

241 'I squeezed past him': Black Book 2, see Shamdasani's Introduction to C. G. Jung, *Liber Novus*, p. 17.

242 still noticed nothing unusual: Ulrich Hoerni in the Preface to *Liber Novus*.

243 almost let himself drown: Tina Keller, interview with G. F. Nameche.

243 The children didn't like Tante Toni: author interviews with the Jung family.

243 Carl left for Ravenna: Bair, *Jung*, suggests that Jung went with Toni Wolff, which would have been callous indeed, but the Jung family archive confirms it was Schmid.

244 'So we are rid of them at last': Kerr, *A Most Dangerous Method*, p. 471.

244 'As a psychiatrist I became worried': McGuire and Hull (eds), *C. G. Jung Speaking*, p. 226.

244 the world itself was going mad: for war fever in Zürich see Guggenheim, *Alles in Allem*, Vol. 2, p. 77.

245 'And be kissed by me': with thanks to Andreas Jung for confirmation.

245 'Finally I understood': McGuire and Hull (eds), *C. G. Jung Speaking*, p. 226.

Chapter 12: The Great War

246 during the First World War: for Zürich during wartime see Guggenhiem, *Alles in Allem*, Vol. 2, p. 84.

246 'In my memory': Baumann, *Zürich*, p. 57 (author's translation).

248 Vladimir Ilyich Ulyanov – Lenin: Baumann, *Zurcher Schlagzeilen*, p. 43.

249 the class struggle: Tanner, *Arbeitsame Patrioten*, p. 699 (author's translation).

252 the first train of French wounded: Guggenheim, *Alles in Allem*, Vol. 2, p. 141.

252 his ideas on polygamy: for Jung on marriage see 'Marriage as a Psychological Relationship' (1925) in *CW17: Development of Personality*, ed. and tr. Gerhard Adler and R. F.C. Hull, Princeton, NJ: Princeton University Press,| 1981.

252 'I think she had many admirers': for Emma's beauty see Ruth Bailey, interview with G. F. Nameche.

253 her feelings of dissociation: based on letter in the Jung family archive.

253 'Dear Gretli': Jaffé, *C. G. Jung*, p. 142.

254 'While I was there': for the mandalas see *MDR*, p. 187.

255 only Toni was able to follow him: for Wolff's importance see Susi Trüb, interview with G. F. Nameche.

255 teetering on the edge: *MDR*, p. 181, and Kerr, *A Most Dangerous Method*, p. 503 for his deterioration in 1915.

255 'Phenomenologically one might classify it': Carl Meier, interview with G. F. Nameche.

255 'creative illness': a term coined by Henri Ellenberger, *The Discovery of the Unconscious: The History and Evolution of Dynamic Psychiatry*, New York: Basic Books, 1970, p. 673; see also Murray Stein, *Jung's Treatment of Christianity: The Psychotherapy of a Religious Tradition*, Wilmette, IL: Chiron Publications, 1985, p. 204, for a pattern throughout Jung's life: 'the crisis would then be followed by an experience of

relief and restoration; and the final step would be an attempt to integrate the good and bad sides of the self, often in symbolic formulation …'.

256 He might tease the maid, flick peas: author interviews with the Jung family.

256 'He doesn't mean it': Agathe Jung, interview with G. F. Nameche.

258 a tailor arrived from London: author interviews with the Jung family.

257 First thing every morning: the details of family life are based on author interviews with the Jung family, and family memoirs, with thanks to Adrian Baumann.

258 'How lucky you English are': Beryl Pogson, *Maurice Nicoll: A Portrait, London*: Vincent Stuart, 1961

258 'Papa was always full of ideas': Agathe Jung, interview with G. F. Nameche.

259 Franz agreed: author interview with Franz's son Andreas Jung, and Donn, *Freud and Jung*, p. 132.

259 'She tried very hard': Susi Trüb, interview with G. F. Nameche.

260 Groma Jung was always there: Agathe Jung, interview with G. F. Nameche.

261 'But I think perhaps': Susi Trüb, interview with G. F. Nameche.

261 'This was an exceptional relationship': Tina Keller, interview with G. F. Nameche.

262 'There isn't the slightest doubt': Donn, *Freud and Jung*, p. 180, quoting Fowler McCormick.

262 'Mona Lisa': see McGuire's Introduction to Jung, *Introduction to Jungian Psychology*, p. xiv.

262 Every now and again Emma blew: author interviews with the Jung family.

263 'dangerous' company: Tina Keller, interview with G. F. Nameche.

Chapter 13: The Americans

265 Edith and Harold McCormick: for the McCormicks see Ron Chernow, *Titan: The Life of John D. Rockefeller, Sr.*, New York: Random House, 1998, p. 414.

266 Harold's cousin Medill: McCormick: McGuire, 'Firm Affinities', p. 305.

267 set him up in America: Shamdasani, *Cult Fictions*, p. 21, and Hannah, *Jung*, p. 109.

267 'Edith is becoming': Chernow, *Titan*, p. 599.

267 a more 'normal' life: Richard Noll, *The Jung Cult: Origins of a Charismatic Movement*, Princeton, NJ: Princeton University Presss, 1994, p. 212.

268 Muriel and Mathilda: Agathe Jung, interview with G. F. Nameche.

268 'Now please look at Mrs Jung': Fowler McCormick, interview with G. F. Nameche.

268 'Her step is springy': Noll, *The Jung Cult*, p. 211.

269 a walking tour: Noll, *The Jung Cult*, p. 216.

266 her hats were marvellous: author interviews with the Jung family, and family memoirs, with thanks to Adrian Baumann.

269 'I am enclosing': Chernow, *Titan*, p. 602.

270 'So Swiss ethics': Shamdasani, *Cult Fictions*, p. 21.

270 'He was too strong': Alphonse Maeder, interview with G. F. Nameche, and Donn, *Freud and Jung*, p. 133.

270 'Jung was very critical': Donn, *Freud and Jung*, p. 126.

271 'vulgar and repellent': Tina Keller, interview with G. F. Nameche.

271 remembered Emma Jung saying: Maria Schmid, interview with G. F. Nameche.

271 The first meeting was held: for the Psychological Club see Friedel Muser, *Zur Geschichte des Psychologischen Clubs Zürich von den Anfängen bis 1928*, Zürich: Psychologischer Club, 1984, and Shamdasani, *Cult Fictions*, p. 43.

272 it was too luxurious: Shamdasani, *Cult Fictions*, p. 43, and Hannah, *Jung*, p. 130.

272 'unconsciously there is too much': for McCorminck's view of the club see Shamdasani, *Cult Fictions*, p. 43.

273 Jung himself was clear: for Jung's view of the club see Shamdasani, *Cult Fictions*, p. 24.

274 'If the Club hasn't as yet': Imelda Gaudissart, *Emma Jung: Analyste et écrivain*, Lausanne: L'Âge d'Homme, 2010, p. 92.

274 Maria Moltzer … Fanny Bowditch: Shamdasani, *Cult Fictions*, p. 66; Moltzer added: 'A club cannot survive unless it is financed by its members … The present members should find it disgraceful to be parasitic.'

275 she chose 'Guilt': Gaudissart, *Emma Jung*, p. 93, and Psychological Club archive.

275 'I'll take her on my knee': Alphonse Maeder, interview with G. F. Nameche.

276 'You must be quite clear': Tina Keller, interview with G. F. Nameche.

276 'My father would not confess it': Donn, *Freud and Jung*, p. 27.

276 'Jung [was] religiously interested': Alphonse Maeder, interview with G. F. Nameche.

276 'religious in a somewhat other sense': For those interested in a summary of the debate based on Noll's *The Jung Cult*, see Bair, *Jung*, p. 741, fn 17, and for an excellent critique of Noll see Shamdasani, *Cult Fictions*.

277 'She had an extraordinary genius': Hannah, *Jung*, p. 118.

277 'It is unfortunately true': Jung, *Letters of C. G. Jung*, Vol. 2, p. 455, and Wehr, *An Illustrated Biography*, p. 189.

277 'She wasn't at all practical': Susi Trüb, interview with G. F. Nameche.

278 'It began with restlessness': *MDR*, p. 182.

279 'As day fades': Jung, *Introduction to Jungian Psychology*, p. 67.

279 'No doubt it was': *MDR*, p. 183.

280 'All summer long': Donn, *Freud and Jung*, p. 173, and author interviews with Jung and Homberger family, and family memoirs, with thanks to Adrian Baumann.

280 Bertha Rauschenbach never worried: author interview with Beatrice Homberger.

280 Many weekends and Easters: the details of the weekends and other holidays and Christmas based on author interviews with the Jung family.

284 By 1917: for Zürich in 1917 see Guggenheim, *Alles in Allem*, Vol. 2, p. 208.

285 there was a general strike: Baumann, *Zürcher Schlagzeilen*, p. 53.

286 back to normal: for the end of the war see Guggenheim, *Alles in Allem*, Vol. 2, p. 288.

286 117 days: see Shamdasani's Introduction to Jung, *Liber Novus*, p. 26.

286 Carl's emergence: *MDR*, p. 186, with thanks to Sonu Shamdasani for confirmation of identity. See Shamdasani's Introduction to Jung, *Liber Novus*, p. 44: 21 November 1918, 'M Moltzer has again disturbed me with letters.'

Chapter 14: Into the Twenties

287 'London, 1 July 1919': Jung, *Letters of C. G. Jung*, Vol. 1, p. 36.

287 Marianne was the musical one: Walter Niehus, interview with G. F. Nameche.

289 Gret's memory: the children's reminiscences of their parents based on Agathe Jung, interview with G. F. Nameche, and author interviews with the Jung family for the use of *Papa* and *Mama*.

289 he didn't mean it: author interview with Carl and Emma's grandson Dieter Baumann.

289 Franz was the 'soft' one: author interview with Franz's son Andreas Jung.

290 Perhaps Lil was the lucky one: for Hélène's memories of Emma see Gaudissart, *Emma Jung*, p. 161, and family memoirs, with thanks to Adrian Baumann.

291 'always the lady': author interview with Beatrice Merk.

291 'Father was often terribly funny': Agathe Jung, interview with G. F. Nameche.

291 'The way he used to sip': Adrian Baumann, interview with G. F. Nameche, and author interview with Adrian Baumann.

292 '*syndrome ambulatoire*': Bair, *Jung*, p. 315.

292 'Jung unfortunately had a great success': McGuire, 'Firm Affinities', p. 312.

292 recounted to Emma in his letters: *MDR*, Appendix, and Jaffé, *C. G. Jung*, p. 150.

294 'At last I was where': *MDR*, p. 225.

296 the Grail legend: see the Introduction to Emma Jung and and Marie-Louise von Franz, *The Grail Legend*, second edition, tr. Andrea Dykes, Princeton, NJ: Princeton University Press, 1998. It is possible Marie-Louise von Franz wrote the Introduction from Emma's notes when Carl asked her to finish the book after Emma's death.

299 'Her influence': Heinrich Fierz, interview with G. F. Nameche.

299 'It wasn't just a triangle situation': Johanna Meier, interview with G. F. Nameche, and Donn, *Freud and Jung*, p. 180.

299 'She was his partner': Tina Keller, interview with G. F. Nameche.

300 a boy was standing: Hans Kuhn, interview with G. F. Nameche.

300 village called Bollingen: the building of the the house at Bollingen and life there based on author interviews with the Jung family and G. F. Nameche interviews.

301 the famous Tower at Bollingen: *MDR*, p. 212.

304 the *Milwaukee Journal*: Bair, *Jung*, p. 341.

305 for Mombasa in Kenya: Ruth Bailey, interview with G. F. Nameche.

307 Daniel Hislop: McGuire and Hull (eds), *C. G. Jung Speaking*, p. 50.

307 a long letter to young Hans Kuhn: Jung, *Letters of C. G. Jung*, Vol. 1, p. 64.

308 Franz, now eighteen, would go: author interview with Franz's son Andreas Jung, and Donn, *Freud and Jung*, p. 181.

308 having her portrait painted: Ruth Bailey, interview with G. F. Nameche.

309 'highly charged': *MDR*, p. 255, and Bair, *Jung*, p. 356.

309 'I can let myself be deceived': *MDR*, p. 60.

309 'For Mrs Jung': Fowler McCormick, interview with G. F. Nameche.

310 'had to be on her husband's side': Susi Trüb, interview with G. F. Nameche.

310 From March to July 1925: Jung, *Introduction to Jungian Psychology*.

311 the Jungian theme of animus/anima: Emma Jung, *Animus and Anima*, tr. Cary F. Baynes and Hildegard Nagel, New York: Spring Publications, 1957, p. 15.

313 'There is no such thing': 'Marriage as a Psychological Relationship' (1925) in Jung, *CW17*.

314 'I went to the train': Jolande Jacobi, interview with G. F. Nameche.

315 'an excellent opportunity': Freud and Jones, *The Complete Correspondence*, p. 640.

Chapter 15: Coming Through

319 'Oh my God': Brome, *Jung*, p. 225.

319 'She was just beginning': Barbara Hannah, interview with G. F. Nameche, and Hannah, *Jung*, p. 201.

319 'She approached the problem': Eleanor Stone in *Bulletin of the Analytical Psychology Club of New York*, Vol. 17, Issue 5,

1955, courtesy of the Kristine Mann Library of the Psychological Club of New York.

320 shouting *Alleluia!*: Anthony, *The Valkyries*, p. 24.

320 'those fawning women': Hayman, *A Life of Jung*, p. 275, and Ferne Jensen (ed.), *C. G. Jung, Emma Jung, and Toni Wolff: A Collection of Remembrances*, San Francisco, CA: Analytical Psychology Club of San Francisco, 1982, p. 101.

320 'Emma was just lovely': Diana Baynes Janson, *Jung's Apprentice: A Biography of Helton Godwin Baynes*, Einsiedeln: Daimon Verlag, 2003, p. 151.

321 'Dear Mr McCormick': with thanks to the Wisconsin Historical Archive.

322 'I think she underwent': Donn, *Freud and Jung*, p. 180, and Carl Meier, interview with G. F. Nameche.

322 this marriage relationship: Joanna Meier, interview with G. F. Nameche.

323 'joint career': thanks to Sonu Shamdasani for his useful term 'joint career' to describe the extent of Carl and Emma's partnership.

323 'I do love Emma': Baynes Jansen, *Jung's Apprentice*, p. 258.

323 Toni was a manipulator': Jane Hollister Wheelwright, interview with Marion Woodman, *Jung Journal*, Volume 5, Issue 1, 2011, courtesy of the Kristine Mann Library of the Psychological Club of New York.

324 Carl went to Berlin: Thomas Kirsch, 'Toni Wolff–James Kirsch Correspondence', *Journal of Analytical Psychology*, Vol. 48, p. 501.

324 'Jung told me personally': Joseph Henderson, interview with G. F. Nameche.

324 'The thing went on': Jane Hollister Wheelwright, interview with Marion Woodman, *Jung Journal*.

324 'Toni overcame': Barbara Hannah, interview with G. F. Nameche.

324 'The law Jung': Carl Meier, interview with G. F. Nameche.

325 'You would then get to know her a bit': Jensen (ed.), *C. G. Jung, Emma Jung, and Toni Wolff*, p. 48. When *MDR* was published many were shocked to find no mention of Toni Wolff, though she was included in the 'Protocols' which form the basis of *MDR*. See Bair, *Jung*, Chapter 38, for a full discussion.

325 'She did not give herself easily': Jensen (ed.), *C. G. Jung, Emma Jung, and Toni Wolff*, p. 1.

325 'My original image': Jensen (ed.), *C. G. Jung, Emma Jung, and Toni Wolff*, p. 10.

326 America in 1936: Hannah, *Jung*, p. 237.

327 back to America: Hannah, *Jung*, p. 238, and McGuire, *Bollingen: An Adventure in Collecting the Past*, Princeton, NJ: Princeton University Press, 1982, p. 12.

327 Mary and Paul Mellon: William McGuire, *Bollingen*, p. 3, and William Schoenl, *C. G. Jung: His Friendships with Mary Mellon and J. B. Priestly*, Wilmette, IL: Chiron Press, 1998, p. 3.

329 perhaps anti-Semitic: for the accusations of anti-Semitism see Shamdasani, *Freud Files*, p. 277, and with thanks to Sonu Shamdasani for letter to Henry Murray, 19 December 1938: 'The origin of the story about myself being a frequent guest a Berchtesgaden has been traced back to Dr Hadley Cantil ... I should very much like to know what on earth prompted this man to tell such a cock-and-bull story ... there must be something behind it ... Whatever I touch and wherever I go I meet with this prejudice that I am a Nazi ... not persecution delusions ... a very real difficulty.'

330 'It was an extremely happy time': Baynes Jansen, *Jung's Apprentice*, p. 285.

331 'My entire youth': *MDR*, p. 52.

332 lead Jung to alchemy: McGuire and Hull (eds), *C. G. Jung Speaking*, p. 221, and *MDR*, p. 189.

332 'She was afraid': Susi Trüb, interview with G. F. Nameche.

332 'That was a terrible situation': Ruth Bailey, interview with G. F. Nameche.

332 'When I was a ripe man': Donn, *Freud and Jung*, p. 181.

332 Fowler McCormick ... noticed: Fowler Mc Cormick, interview with G. F. Nameche.

333 'Well, she was a marvellous woman': Jolanda Jacobi, interview with G. F. Nameche.

333 'I think the night': McGuire, *Bollingen*, pp. 34, 49.

334 Emma who enforced it: for Emma's guarding of Carl see Susi Trüb, interview with G. F. Nameche, and Ruth Bailey, interview with G. F. Nameche.

336 'She was the heart': author interviews with Adrian Baumann and Andreas Jung, and Hans and Jost Horni, interview with G. F. Nameche.

336 'It's remarkable how': author interview with Dieter Baumann.

336 Emma was clever: author interview with Brigitte Merk.

336 'Oh she was la Grande Dame': Ania Teillard, interview with G. F. Nameche.

336 nineteen grandchildren: Author interviews with Adrian, Klaus and Dieter Baumann.

338 the Jung Institute was founded: Jensen (ed.), *C. G. Jung, Emma Jung, and Toni Wolff*, pp. 40–41.

339 'Lotus. Nun. Mysterious': Bair, *Jung*, p. 559; this is a loose translation, but makes the point.

339 'I shall always be grateful': Hannah, *Jung*.

340 'You have done the right thing': Sabi Tauber, interview with G. F. Nameche. See McGuire and Hull (eds), *C. G. Jung Speaking*, p. 220, for Jung's Answer to Job interpretation regarding his father.

340 danced a waltz: Jost Hoerni, interview with G. F. Nameche.

340 'all tense and white': Ruth Bailey, interview with G. F. Nameche.

340 During those last few days: author interviews with Andreas Jung and Adrian Baumann.

341 'I cannot continue': for Jung's letters following Emma's death see Jung, *Letters of C. G. Jung*, Vol. 2, pp. 279, 282, 284 and 316.

342 'We went to see him': Frieda Fordham, interview with G. F. Nameche.

342 'He was sad': Ruth Bailey, interview with G. F. Nameche.

342 'I know that when she died': author interview with Adrian Baumann.

342 'Mama's death': Jung, *Letters of C. G. Jung*, Vol. 2, p. 316.

342 'Let every man': Reverend Hans Schaer's funeral address courtesy of the Kristine Mann Library of the Psychological Club of New York.

344 'I saw her in a dream': *MDR*, p. 276.

Bibliography

Anthony, Maggy, *The Valkyries: The Women Around Jung*, Longmead: Element Books, 1990

Appignanesi, Lisa, and John Forrester, *Freud's Women*, London: Weidenfeld and Nicolson, 1992

Arp, Hans, *Zürcher Erinnerungen aus der Zeit des Ersten Weltkrieges*, Zurich: Artemis Verlag, 1954

Astor, James, 'Our Cause', *British Journal of Psychotherapy*, 11 July 2008

Bair, Deirdre, *Jung: A Biography*, Boston, MA: Little, Brown, 2003

Baumann, Walter, *Zürich: La Belle Époque*, Zürich: Orell Füssli Verlag, 1973

—, *Zürcher Schlagzeilen*, Zürich: Orell Füssli Verlag, 1981

Baynes Jansen, Diana, *Jung's Apprentice: A Biography of Helton Godwin Baynes*, Zürich: Daimon Verlag, 2003

Behling, Katja, *Martha Freud: A Biography*, Cambridge: Polity Press, 2005

Bennet, E. A., *C. G. Jung*, London: Barrie & Rockliff, 1961

—, *Meetings with Jung*, Zürich: Daimon Verlag, 1985

Bennet, Glin, 'Domestic Life with C. G. Jung: Tape-recorded Conversations with Ruth Bailey', *Spring*, 1986: pp. 177–89

Binswanger, Ludwig, *Sigmund Freud: Reminiscences of a Friendship*, tr. Norbert Guterman, New York: Grune & Stratton, 1957

Borch-Jacobsen, Mikkel, and Shamdasani, Sonu, *The Freud Files: An Inquiry into the History of Psychoanalysis*, Cambridge: Cambridge University Press, 2012

Brill, Abraham, *Lectures on Psychoanalytic Psychiatry*, New York: Knopf, 1946

Brome, Vincent, *Jung: Man and Myth*, London: Macmillan, 1978

—, *Ernest Jones: Freud's Alter Ego*, London: Caliban Books, 1982

Brown, Kevin, *The Pox: The Life and Near Death of a Very Social Disease*, Stroud: Sutton Publishing, 2006

Carotenuto, Aldo, *A Secret Symmetry: Sabina Spielrein Between Jung and Freud*, New York: Pantheon Books, 1982

Champernowne, Irene, *A Memoir of Toni Wolff*, San Francisco: C. G. Jung Institute of San Francisco, 1980

Chernow, Ron, *Titan: The Life of John D. Rockefeller, Sr.*, New York: Random House, 1998

Clay, Catrine, *King, Kaiser, Tsar: Three Royal Cousins Who Led the World to War*, London: John Murray, 2006

Covington, Coline, and Barbara Wharton (eds), *Sabina Spielrein: Forgotten Pioneer of Psychoanalysis*, London: Brunner-Routledge, 2003

Danuser, Hanspeter, *Das Trambuch: 100 Jahre Züri-Tram*, Zürich: Verlag Neue Zürcher Zeitung, 1982

Donn, Linda, *Freud and Jung: Years of Friendship, Years of Loss*, New York: Charles Scribner's Sons, 1988

Dunne, Claire, *Carl Jung: Wounded Healer of the Soul*, New York: Parabola Books, 2000

Dyer, Donald R., *Cross-Currents of Jungian Thought: An Annotated Bibliography*, Boston: Shambhala Publications, 1991

Ellenberger, Henri, *The Discovery of the Unconscious: The History and Evolution of Dynamic Psychiatry*, New York: Basic Books, 1970

Elms, Alan C., *Uncovering Lives: The Uneasy Alliance of Biography and Psychology*, Oxford: Oxford University Press, 1994

Ferris, Paul, *Dr Freud: A Life*, London: Sinclair-Stevenson, 1997

Fordham, Michael, 'Memories and Thoughts about C. G. Jung', *Journal of Analytical Psychology*, Vol. 20, Issue 2, July 1975

Freud, Ernst, Lucie Freud and Ilse Grubrich-Simitis (eds), *Sigmund Freud: His Life in Pictures and Words*, tr. Christine Trollope, London: André Deutsch, 1978

Freud, Martin, *Glory Reflected: Sigmund Freud – Man and Father*, London: Angus & Robertson, 1957

Freud, Sigmund, *The Interpretation of Dreams* in *The Standard Edition of the Complete Psychological Works of Sigmund Freud*, Vols 4–5, New York: W. W. Norton, 2000

—, *The Complete Letters of Sigmund Freud to Wilhelm Fliess, 1887–1904*, tr. and ed. Jeffrey Moussaieff Masson, Cambridge, MA: Harvard University Press, 1985

—, *Unser Herz zeigt nach dem Süden. Reisebriefe 1895–1923*, ed. Christfried Tögel, Berlin: Aufbau Verlag, 2002

— and Karl Abraham, *Sigmund Freud–Karl Abraham: Briefe 1907–1926*, ed. Hilda C. Abraham and Ernst L. Freud, Frankfurt: S. Fischer Verlag, 1965

—, *The Complete Correspondence of Sigmund Freud and Karl Abraham, 1907–1925*, ed. Ernst Falzeder, London: Karnac Books, 2002

— and Ludwig Binswanger, *Sigmund Freud–Ludwig Binswanger, Briefwechsel 1908–1938*, ed. Gerhard Fichtner, Frankfurt: S. Fischer Verlag, 1992

— and Sándor Ferenczi, *The Correspondence of Sigmund Freud and Sándor Ferenczi, Volume 1: 1908–1914*, ed. Eva Brabant, Ernst Falzeder and Patrizia Giampieri-Deutsch, tr. Peter T. Hoffer, Cambridge, MA: Harvard University Press, 1993

— and Ernest Jones, *The Complete Correspondence of Sigmund Freud and Ernest Jones, 1908–1939*, ed. R. Andrew Paskauskas, Cambridge, MA: Harvard University Press, 1993

— and C. G. Jung, *The Freud/Jung Letters: The Correspondence between Sigmund Freud and C. G. Jung*, ed. William McGuire, tr. Ralph Manheim and R. F. C. Hull, Princeton, NJ: Princeton University Press, 1974

Gaudissart, Imelda, *Emma Jung: Analyste et écrivain*, Lausanne: L'Âge d'Homme, 2010

Graf-Nold, Angela, 'The Zürich School of Psychiatry in Theory and Practice: Sabina Spielrein's Treatment at the Burghölzli Clinic in Zürich', *Journal of Analytical Psychology*, Vol. 46, Issue 1, January 2001

—, '100 Jahre Peinlichkeit – und (k)ein Ende?: Karl Abrahams frühe psychoanalytische Veröffentlichungen und die "sexuelle Frage"', *Jahrbuch der Psychoanalyse* 52 (2006): 93–138

Guggenheim, Kurt, *Alles in Allem*, 2 vols, Zürich: Artemis Verlag, 1954

Hannah, Barbara, *Jung: His Life and Work*, London: Michael Joseph, 1976

Hayman, Ronald, *A Life of Jung*, London: Bloomsbury, 1999

Jacobi, Jolanda, *The Psychology of C. G. Jung*, New Haven: Yale University Press, 1973

Jaffé, Aniela, *C. G. Jung: Word and Image*, Bollingen Series XCVII:2, Princeton, NJ: Princeton University Press, 1979

James, William, *The Letters of William James*, ed. Henry James, London: Longmans, Green, and Co., 1920

Janik, Allan, and Stephen Toulmin, *Wittgenstein's Vienna*, London: Weidenfeld and Nicolson, 1973

Jensen, Ferne (ed.), *C. G. Jung, Emma Jung, and Toni Wolff: A Collection of Remembrances*, San Francisco, CA: Analytical Psychology Club of San Francisco, 1982

Jones, Ernest, *The Life and Work of Sigmund Freud*, 3 vols, London: Hogarth Press, 1953–7

Jung, Andreas, Regula Michel, Arthur Ruegg, Judith Rohrer and Daniel Ganz, *The House of C. G. Jung: The History and Restoration of the Residence of Emma and Carl Gustav Jung-Rauschenbach*, Wilmette, IL: Chiron Publications, 2008

Jung, C. G., *Collected Works of C. G. Jung*, Vol. 1, *Psychiatric Studies*, ed. Herbert Read, Michael Fordham and Gerard Adler, tr. R. F. C. Hull, Princeton, NJ: Princeton University Press, 1970

—, *Studies in Word Association*, in *Collected Works of C. G. Jung*, Vol. 2, *Experimental Researches*, ed. and tr. Gerhard Adler and R. F. C. Hull, Princeton, NJ: Princeton University Press, 1973

—, *On the Psychology of Dementia Praecox*, in *Collected Works of C. G. Jung*, Vol. 3, *Psychogenesis of Mental Disease*, ed. and tr. Herbert Read, R. F. C. Hull and Gerhard Adler, Princeton, NJ: Princeton University Press, 1961

—, *Collected Works of C. G. Jung*, Vol. 5, *Symbols of Transformation*, ed. and tr. Gerhard Adler and R. F.C. Hull, Princeton, NJ: Princeton University Press, 1977

—, *Collected Works of C. G. Jung*, Vol. 17, *Development of Personality*, ed. and tr. Gerhard Adler and R. F.C. Hull, Princeton, NJ: Princeton University Press,| 1981

—, *Collected Works of C. G. Jung*, Vol. 18, *The Symbolic Life: Miscellaneous Writings*, ed. and tr. Gerhard Adler and R. F. C. Hull, Princeton, NJ: Princeton University Press, 1977

—, *Letters of C.G. Jung*, 2 vols, ed. Gerhard Adler and Aniela Jaffé, tr. R. F. C. Hull, Princeton, NJ: Princeton University Press, 1973–75

—, *Introduction to Jungian Psychology: Notes of the Seminar on Analytical Psychology Given in 1925*, ed. William McGuire, tr. R F. C. Hull, new Introduction and updates by Sonu Shamdasani, Princeton, NJ: Princeton University Press, 1989

—, *The Red Book: Liber Novus*, ed. Sonu Shamdasani, tr. M. Kyburz, J. Peck and S. Shamdasani, New York: W. W. Norton & Company, 2009

— and Jaffé, Aniela, *Memories, Dreams, Reflections*, New York: Pantheon, 1962

Jung, Emma, *Animus and Anima*, tr. Cary F. Baynes and Hildegard Nagel, New York: Spring Publications, 1957

— and Marie-Louise von Franz, *The Grail Legend*, second edition, tr. Andrea Dykes, Princeton, NJ: Princeton University Press, 1998

Kerr, John, *A Most Dangerous Method: The Story of Jung, Freud, and Sabina Spielrein*, New York: Knopf, 1993

McGuire, William, 'Firm Affinities: Jung's Relations with Britain and the United States', *Journal of Analytical Psychology*, 40.3, July 1995

—, *Bollingen: An Adventure in Collecting the Past*, Princeton, NJ: Princeton University Press, 1982

— and R. F. C. Hull (eds), *C. G. Jung Speaking: Interviews and Encounters*, London: Picador, 1978

Maines, Rachel P., *The Technology of Orgasm: 'Hyesteria', the Vibrator, and Women's Sexual Satisfaction*, Baltimore, MD: Johns Hopkins University Press, 1999

Makari, George, *Revolution in Mind: The Creation of Psychoanalysis*, London: Duckworth, 2008

Minder, Bernard, 'Sabina Spielrein: Jung's Patient at the Burghölzli', *Journal of Analytical Psychology*, Vol. 46, Issue 1, January 2001

Mösli, Rolf, *Eugen Bleuler: Pionier der Psychiatrie*, Zürich: Römerhof Verlag, 2012

Muser, Friedel, *Zur Geschichte des Psychologischen Clubs Zürich von den Anfängen bis 1928*, Zürich: Psychologischer Club, 1984

Noll, Richard, *The Jung Cult: Origins of a Charismatic Movement*, Princeton, NJ: Princeton University Presss, 1994

—, *The Aryan Christ: The Secret Life of Carl Jung*, New York: Random House, 1997

Pogson, Beryl, *Maurice Nicoll: A Portrait, London*: Vincent Stuart, 1961

Quétel, Claude, *History of Syphilis*, tr. Judith Braddock and Brian Pike, Cambridge: Polity Press, 1990

Schoenl, William, *C. G. Jung: His Friendships with Mary Mellon and J. B. Priestly*, Wilmette, IL: Chiron Press, 1998

Schorske, Carl E., *Fin-de-Siècle Vienna: Politics and Culture*, New York: Random House, 1981

Segaller, Stephen and Berger, Merrill, *Jung: The Wisdom of the Dream*, London: Weidenfeld & Nicolson, 1989

Shamdasani, Sonu, *Cult Fictions: C. G. Jung and the Founding of Analytical Psychology*, London: Routledge, 1998

—, *Jung Stripped Bare: By His Biographers, Even*, London: Karnac Books, 2005

—, *C. G. Jung: A Biography in Books*, New York: W. W. Norton & Company, 2012

Showalter, Elaine (ed.), *These Modern Women: Autobiographical Essays from the Twenties*, New York: The Feminist Press, 2003

Stein, Murray, *Jung's Treatment of Christianity: The Psychotherapy of a Religious Tradition*, Wilmette, IL: Chiron Publications, 1985

Tanner, Albert, *Arbeitsame Patrioten – wohlanständige Damen: Bürgertum und Bürgerlichkeit in der Schweiz, 1830–1914*, Zürich: Orell Füssli Verlag, 1995

Taylor, A. J. P., *The Habsburg Monarchy, 1809–1918: A History of the Austrian Empire and Austria-Hungary*, London: Hamish Hamilton, 1948

Waissenberger, Robert (ed.), *Traum und Wirklichkeit: Wien 1870–1930*, Vienna: Residenz Verlag, 1984

Wehr, Gerhard, *An Illustrated Biography of C. G. Jung*, tr. Michael Kohn, Boston: Shambhala Publications, 1989

—, *Jung: A Biography*, tr. David M. Weeks, Boston: Shambhala Publications, 1987

Weissweiler, Eva, *Die Freuds: Biographie einer Familie*, Cologne: Kiepenhauer & Witsch, 2006

Zweig, Stefan, *Die Welt von Gestern: Erinnerungen eines Europäers*, Stockholm: Bermann-Fischer Verlag, 1941

Credits

Unless otherwise stated, the images belong to the Jung Family Archive. They fall into two categories: those previously published elsewhere, and those never previously published. For the latter I wish to thank Andreas Jung for his generous help. These include: Emma aged sixteen; Emma at school; Emma in Canton of Schaffhausen costume; Emma and Carl's engagement; Emma on her wedding day; Carl with Aggie and Gretli; Emma with Gretli; Emma with Aggie and Gretli outside Burgholzli; the studio portrait of Emma; Emma and Toni at the Psychological Club; Emma and Carl at Eranos in the 1930s.

Section One

Page 2 (bottom): Vienna c. 1907, Stadtsarchiv Vienna

Page 3: Emma and Marguerite, with thanks to Beatrice Homberger

Page 4 (top left): The Haus zum Rosengarten, Staatsarchiv Schaffhausen

Page 4 (top right): The Rhine Falls, Staatsarchiv Schaffhausen/ Fotosammlung Neuhausen am Rheinfall

Page 6 (top left): Jean Rauschenbach, with thanks to Beatrice Homberger

Page 8 (top): Burghölzli Asylum, with thanks to Jorg Zemp,
  Burghölzli Archive
Page 8 (bottom): men's wards at Burghölzli, with thanks to
  Jorg Zemp, Burghölzli Archive

Section Two
Page 2 (bottom): Zürich c. 1905, Schweizerische
  Nationalbibliothek/Eidgenossisches Denkmalpflege
Page 7 (top): Toni Wolff, collection Jane and Joseph
  Wheelwright
Page 7 (bottom left): Carl and Toni Wolff in England, 1922,
  courtesy of the Kristine Mann Library, New York

Integrated
Page 130: the galvanometer, Aniela Jaffé, *C. G. Jung: Word and
  Image*
Page 273: Harold McCormick with his two daughters,
  Wisconsin Historical Society, WHS-114385
Page 283: Bertha Rauschenbach, with thanks to Beatrice
  Homberger
Page 328: Mary and Paul Mellon, Eranos Foundation

The author and publishers are committed to respecting the intellectual property rights of others and have made all reasonable efforts to trace the copyright owners of the images reproduced, and to provide appropriate acknowledgement within this book. In the event that any untraceable copyright owners come forward after the publication of this book, the author and publishers will use all reasonable endeavours to rectify the position accordingly.

Acknowledgements

When I first thought of Emma Jung as a subject for my next book, I rang my Swiss cousin Gaby in Zurich. I am half Swiss. Gaby and I spent every summer of our childhood together in the family home in St Gallen, an hour east of Zürich. I rang her from London: could she contact the Jung Institute for me and perhaps arrange a visit to the Jung house at Kusnacht on Lake Zürich? I knew it wouldn't be easy because the Jung family are very private people. But in her quiet way Gaby was persuasive. On my next trip to Zürich we were able to visit the Jung house, shown around by Andreas Jung, Emma and Carl's grandson, who lives there with his wife Vreni.

I thank Andreas and Vreni Jung for their invaluable help and their friendship. Equally, all the other surviving Jung grandchildren who trusted me with their grandmother's remarkable story, passing me from one cousin to the next. They shared one thing: they loved and admired their grandmother and felt that the Jung story, so often told, always neglected her, the key to it all. Or misrepresented her, which was worse. Without the vivid memories of the grandchildren I could not have written this family-based book. A cousin of the cousins, Beatrice Homberger, has my special thanks for furnishing me with some wonderful family photographs along with the many engaging talks we had in Schaffhausen together with Brigitte Merk, another Jung grandchild. A special thanks to Adrian Baumann, another grandchild, who gave up many hours to talk about his beloved grandmother. And to Thomas Fischer, Director of the Foundation of the Works of C. G. Jung, for proper clarification of the terms of the Jung archive.

Nicola Behrens at the Stadtarchiv Zürich helped me to locate newspapers, journals and maps of Zürich at the beginning of the twentieth century. And he was a marvel at finding out about the weather on any particular day: snow, rain, fog, hot summer sun. At Stadtarchiv Schaffhausen Frau Zimmermann and her team were equally helpful and friendly. Angela Graf-Nold, a Jungian who has worked extensively in the Burgholzli asylum archives in Zürich, was an excellent sounding board. Rolf Mosli is the man who started that archive, and he pointed me to some useful sources, not least his own excellent book on Eugen Bleuler. Murray Stein, an American Jungian living in Zürich, offered some interesting interpretations. In Vienna I had the help of a keen young researcher, Valentin Freyler. I thank Ingrid von Rosenberg for putting us in touch, and for giving me some good suggestions about women's lives in Germany and Switzerland at that time. In America I owe thanks to Jack Ebert at the Countway Library at the Harvard Medical School in Boston. Thanks also to the Kristine Mann Library in New York for stories of American friends and patients of the Jungs. And to David Williams for research into the fascinating McCormick/ Rockefeller story at the Wisconsin Historical Society.

In England, Sonu Shamdasani, the foremost Jungian scholar, offered many useful ideas, most especially that the relationship between Carl and Emma Jung was by way of a 'joint career'. Richard Cohen and James Astor each made some excellent suggestions. The staff at the London Library were as helpful as ever in locating long out-of-print books and erudite journals.

Arabella Pike at William Collins has been the kind of editor any writer might appreciate, making strong comments with a nice ease, happily supported by Stephen Guise, senior project editor, and Kate Johnson, the copy-editor. Jo Walker has designed a clever and beautiful jacket. As for my long-established literary agent, Anthony Sheil, the book would not have been written without him. 'Say what you mean, and mean what you say' is his mantra, and now mine. If thanks to my husband John comes last, that's typical of marriage, as Emma Jung could have told him. But it doesn't stop him being central to this book, helping with the research, reading every chapter, offering diplomatic criticisms.

All the time I was writing the book Gaby was suffering from cancer. She died last year. The book was almost finished by then and she'd been part of it all along, understanding exactly why I wanted to write it, because it was just the kind of thing we'd always talked about during those long summer holidays: family relationships, complicated marriages, women's lives, happiness and unhappiness, ambition, loyalty, love, laughter. The book is dedicated to her, with my deepest thanks.

Index